NEBS
MANAGEMENT
DEVELOPMENT

SUPER SERIES

THIRD EDITION
Managing Activities

Managing a Safe Environment

Published for
&NEBS Management *by*

Pergamon
Open
Learning

Pergamon Open Learning
An imprint of Butterworth-Heinemann
Linacre House, Jordan Hill, Oxford OX2 8DP
A division of Reed Educational and Professional Publishing Ltd

ℛ A member of the Reed Elsevier plc group

OXFORD   BOSTON   JOHANNESBURG
MELBOURNE   NEW DELHI   SINGAPORE

First published 1986
Second edition 1991
Third edition 1997

British Library Cataloguing in Publication Data
A catalogue record for this book is available from the British Library

ISBN 0 7506 3300 X

NEBS Management Project Manager: Diana Thomas
Author: Joe Johnson
Editor: Fiona Carey
Series Editor: Diana Thomas
Based on previous material by: Joe Johnson
Composition by Genesis Typesetting, Rochester, Kent
Printed and bound in Great Britain

Contents

Performance checks   73

Reflect and review   79

Workbook introduction

1 NEBS Management Super Series 3 study links

Here are the workbook titles in each module which link with *Managing a Safe Environment*, should you wish to extend your study to other Super Series workbooks. There is a brief description of each workbook in the User Guide.

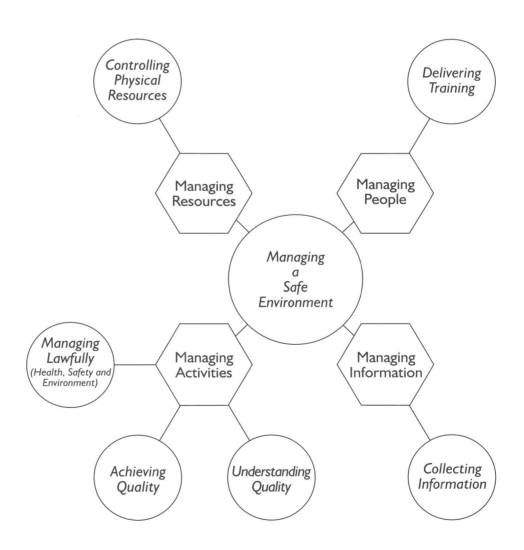

2 S/NVQ links

This workbook relates to the following elements:

B1.2 Contribute to the control of resources
A1.2 Maintain healthy, safe and productive working conditions

It is designed to help you to demonstrate the following Personal Competences:

- building teams;
- focusing on results;
- thinking and taking decisions;
- striving for excellence.

3 Workbook objectives

EXTENSION 1
This book is listed on page 83.

'The environmental debate is over. Being green is no longer fashionable or glamorous, it is a new fact of business life. Just as the debate about quality and health and safety several years ago established these factors as cornerstones of effective business practice, so environmental issues are here to stay.'
– *Successful Environmental Management in a Week*, by Mark Yoxon.[1]

You may care passionately about the environment, or you may cover your ears whenever you hear the word mentioned. But managers these days can't escape the fact that environmental concerns, just like quality, and health and safety, have to be taken into account in all their decision making.

This is a comparatively new state of affairs, coming about during the past twenty years or so. It is the result of a growing awareness of ecological issues by the general public, who have been bombarded with information by the media.

The good news is that, for business organizations, there are opportunities as well as challenges in this situation.

In Session A of this workbook, we will review some of the major ecological concerns, including: acid rain; ozone depletion; global warming; human population growth; pollution; noise; and energy conservation.

[1] Headway. Hodder and Stoughton, 1996. Page 6. 338 Euston Road, London NW1 3BH.

Session B is entitled 'Responding to the problems', and covers: an introduction to environmental law; public opinion; the environmentalists; actions that could be taken to improve the environment; and discusses what the responsibilities of organizations and individuals should be.

Sessions C and D are concerned with the roles that work organizations and the individuals within them have to play.

3.1 Objectives

When you have completed this workbook you will be better able to:

- have an appreciation of many of the world's ecological problems, and recognize their causes;
- summarize the response to these problems by governments, environmentalists, the general public, the media, and other groups;
- explain why the environment is important to work organizations, and say what kind of actions are appropriate for them to take;
- take a number of positive steps in your own work area to protect and enhance the local and global environment.

4 Activity planner

You may want to look at the following Activities now, so that you can start collecting material as soon as possible:

- To complete Activity 24 you will need to get hold of a copy of your (or another) organization's environment policy document (page 52).
- In Activity 26 you are asked to think about the areas and aspects of your organization's (or department's) activities that might be looked into by an environmental auditor (page 55).
- Activity 29 asks you to plan some specific environmental actions (page 62).
- Activity 30 asks you to organize an energy audit of your work area (page 64).
- For Activity 32, you are expected to identify those materials that your organization recycles, and to make suggestions for possible further candidates for recycling (page 67).

Portfolio of evidence

Some or all of these Activities may provide the basis of evidence for your S/NVQ portfolio. All Portfolio Activities and the Work-based assignment are signposted with this icon.

The icon states the elements to which the Portfolio Activities and Work-based assignment relate.

In the Work-based assignment you are required to check that the work conditions under your control comply with the law.

This task is designed to help you meet element A1.2 of the MCI Management Standards: 'Maintain healthy, safe and productive working conditions', 'You may want to prepare for the assignment in advance.

Session A Environmental concerns

1 Introduction

'1995 was a record-breaking year:

- the highest average temperature since records began;
- the lowest levels of reserve grain stocks ever recorded;
- the highest global income per person of all time;
- the highest ever level of fuel-burning emissions;
- the biggest rise in the level of HIV infection;
- the greatest amount of wind generated power.'

– From *Vital Signs 1996–97*, the Worldwatch Institute.[2]

EXTENSION 2
This book is listed on page 83.

Over the past few years it has been very difficult to avoid hearing about 'the environment'. It seems that every time we read a newspaper, watch television or listen to the radio, the subject comes up. People in the media seem to be continually discussing environmental topics such as 'global warming', 'acid rain' and so on, or else highlighting pollution problems such as, for example, fish being killed by industrial effluents discharged into rivers.

We can't seem to avoid the subject, and you may believe we should not try to. There is no doubt that the environment is extremely important, because, in the broadest sense, it is simply a word for the world in which we live. If we don't take care of our planet, it will inevitably become a worse place for us and our fellow creatures to live in. Should the worst come to the worst, humanity won't survive.

In this session, we begin by getting our definitions right, and in particular coming to an agreement about what we mean by the word 'environment'.

EXTENSION 3
If you have access to a computer with a CD-ROM drive, you may want to take a look at the encyclopaedia *Encarta 97* (or a later edition). It has many interesting and useful articles about the environment.

Then we go on to discuss some of the global ecological problems: natural hazards, acid rain, ozone depletion, global warming, and population growth. The next subject is pollution – air pollution, water pollution, the effects of chemicals, and noise pollution. The last topic in this session is energy conservation.

At this stage of the workbook, this range of issues may not all seem directly relevant to your work or your job. But, as we will discuss in later sessions, 'managing the environment' is an important and necessary activity for the modern manager. This session is intended to provide some of the background information. You should read the material carefully, although you can't be expected to remember all the detail.

[2] © Worldwatch Institute, back page. Published by Earthscan Publications Ltd, 1996. Address: 120 Pentonville Road, London N1 9JN. Tel: 0171 278 0433.

2 Definitions

'Environment' is one of those words which can mean anything – or nothing. We'll start by pinning down this elusive idea.

Activity 1

Tick every statement that you believe to be a good definition of the word 'environment', or else give your own definition. The environment is:

a all the things around us; ☐

b the social, physical and cultural conditions affecting the life of an individual or an organization; ☐

c the atmosphere, the earth, the rivers and oceans; ☐

d the problems of world pollution; ☐

e the habitat for all living creatures. ☐

Your definition:

_____

_____

_____

_____

'Ecology' is the study of:

a) the relationships of organisms to one another and to their surroundings;

b) the interaction of people with their physical environment.

You may have decided to tick all of these statements, because they all have relevance to the environment. One dictionary definition is:

'1 External conditions or surroundings, especially those in which people live or work.

2 Ecology: the external surroundings in which a plant or animal lives, which tend to influence its development and behaviour'

(Collins English Dictionary)

The Environmental Protection Act 1990 (EPA) defined the environment as consisting of:

'. . . all, or any of the following media, namely, the air, water and land'

In the Activity above, none of the answers given are wrong. The simplest and most comprehensive definition is (a) 'all the things around us'. Perhaps your own definition was even more illuminating.

We can think of the environment in terms of both **what we can influence**, and **what influences us**. Our actions tend to affect our local environment first, and the people closest to us. But most things we do also have an effect on the rest of the planet; simply by breathing we alter the atmosphere, by taking in oxygen and emitting carbon dioxide!

In turn, other people, and most natural phenomena, impinge on our lives in all kinds of ways. To give some examples:

- if a doctor in (say) Australia finds a cure for a disease, it may save the lives of many in this country;
- a large earthquake occurring in (say) the Philippines could have a profound, if temporary, impact on the climate in the rest of the world;
- a political coup in a far-off state could result in a world shortage of some vital raw material – such as oil;
- an economic recession in any major country (such as the United States) is likely to have knock-on effects in other parts of the world.

In this workbook, we'll look at the environment from both perspectives: how things affect us, and how we affect them. Also, we'll consider the local work environment and the global environment.

3 Major ecological issues

Let us start by identifying some of the major ecological occurrences which affect the environment.

3.1 Natural hazards

Shortly, we will discuss some of the environmental threats that have arisen partly or wholly as a result of the activities of humans. However, we should first remind ourselves of the many **natural** occurrences that can result in disaster and devastation.

Activity 2

2 mins

An earthquake can have terrible consequences. Jot down the names of **two or three** other large-scale natural hazards.

Apart from earthquakes, you might have mentioned volcanoes, floods, tropical storms, droughts and landslides. The world is not an entirely safe place to live, as the following list of facts illustrates.

- **Earthquakes** occur frequently, and usually result in large-scale destruction and loss of life. (It is said that 'Earthquakes don't kill people; buildings do.') A few examples are listed below.

  - On 19 September 1985, an earthquake in Mexico City killed around 25,000 people, and cost the equivalent of some 4 billion dollars in damage and disruption.
  - In October 1989, a 'quake struck south of San Francisco; 62 people lost their lives, and the total repair bill was at least $6 billion.
  - A shock of the same magnitude as the San Francisco 'quake struck Leniakan in Armenia in 1988. Here, however, the effects were much more serious, and more than half the structures in the city were levelled to the ground; 25,000 people died.
  - On 17 January, 1995, an earthquake hit near the city of Kobe. Some 5,000 people died and more than 21,000 were injured. Over 30,000 buildings were damaged by the quake and resulting fires.

  It is worth noting that, although an earthquake cannot be prevented, it can be predicted. The activities of mankind therefore have an influence on the effects of this natural phenomenon, where it occurs. In wealthy countries such as the USA, governments can afford to enforce building and other regulations that minimize the damage, whereas this is less likely to happen in 'third world' economies, such as Mexico and Armenia.

- There are about 500 active **volcanoes** in the world. Volcanic eruption usually causes molten rock to be poured over the surrounding area, and dust and gas to be blown up into the atmosphere. On 18 May 1980, the Mount St. Helens volcano in Washington, USA flattened half a million trees over a radius of 25 km. On 15 June 1991, Mount Pinatubo in the Philippines erupted, and flows advanced 9½ miles down its sides.

- More than 200,000 people world-wide died as a result of **flooding**, in the period 1973 to 1992.

  The activities of mankind may help to bring about flooding, in that some floods are the result of deforestation. Also, the effects of this natural occurrence depend largely on how well the affected area is equipped to deal with it. Recently some very serious floods in the Netherlands were contained and the damage limited, while the same levels of flooding in, say, Bangladesh, 'routinely' cost thousands of lives.

- The most powerful **hurricane** ever recorded occurred in 1988, causing a national disaster to be declared in Jamaica.

- Many thousands of people are killed by **landslides** and **avalanches**.

Natural hazards like these are largely unavoidable. But there are plenty of hazards that are the consequences of human activity. Let's look now at three major ecological problems facing the world:

- acid rain;
- ozone depletion; and
- global warming.

Deforestation: between 1980 and 1990, the world lost an average of 9.95 million hectares of forest area each year, as trees were cut down, and the land put to other uses.

3.2 Acid rain

Fossil fuel: a natural fuel such as coal or gas formed in the geological past from the remains of living organisms.

Acid rain is a term used to describe the fall-out of industrial pollutants, mainly sulphur dioxide and nitrogen dioxide. These chemicals are the result of the burning of fossil fuels, especially coal, and the smelting of sulphide ores. In the atmosphere they react with water and sunlight, and fall to the earth as dry particles or as acidified rain, snow or fog, sometimes several thousand miles from their source. Because acid rain may take the form of rain or dry particles, it is more properly called 'acid deposition'.

Acid rain is not a new phenomenon; it has existed ever since the first industrial plants first gave off noxious fumes.

It has been estimated that the cost, so far, of repairing damage to buildings in the UK as a result of acid rain is around £17 billion.

The effects of acid rain are to be seen in Northern Europe, parts of the USA and Canada, Brazil, India and China. Acid rain has made many lakes unfit for fish, and eats away the surface of buildings. It is also blamed for destroying forests and woodlands, although some scientists would dispute this.

In Western Europe and the USA, the power generation industry is being compelled to reduce its emissions of sulphur dioxide, gradually year by year. The two ways it can do this are by burning coal with a lower sulphur content, and by fitting flue gas desulphurization equipment between the boiler and the chimney. The worst culprits, however, have been the power stations and industries of Eastern Europe.

3.3 Ozone depletion

The ozone layer surrounds the earth in the stratosphere. The stratosphere is the upper part of the atmosphere. It starts about 11 kilometres above the earth's surface and is some 40 km deep, as shown in the figure below.

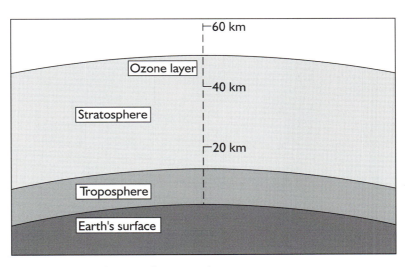

The earth's atmosphere showing the ozone layer

Ozone is a modified form of oxygen found in the upper part of the stratosphere; it acts as a filter to ultra-violet radiation from the sun. Without this filter, normal human, animal and plant life would not be possible. If the ozone becomes depleted, human health could suffer, because of increased incidences of skin cancer, eye cataracts and premature ageing. Sources of food could become seriously affected.

A hole in the ozone layer above Antarctica was first observed in 1975. In the following ten years measurements of ozone levels in this region showed that the levels had decreased by about 50 per cent. This depletion of ozone (occurring at both the South and North Poles) has largely been caused by man-made substances. Chemicals called chlorofluorocarbons (CFCs) (used as propellant gases in aerosol cans, in refrigerators and in plastic food packaging), and halon gases (used in fire extinguishers), react in the atmosphere with ultra-violet radiation to form chlorine. This results in ozone being converted to oxygen, which doesn't have the same filtering properties as ozone.

Incredibly, there is now a black market in CFCs. At least 10,000 tons of these gases were smuggled into the USA in 1995, according to the Worldwatch Institute.

Both the USA and the European Union have agreed to ban CFCs. Although world production has decreased from 1,260 thousand tons in 1988 to around 300 thousand tons in 1995, these gases are so stable they can last for 100 years in the atmosphere. The ozone levels over Antarctica, measured in Dobson units, have diminished considerably since 1979, as the graph below shows:[3]

There are ozone holes over Britain, too, and these are getting larger. The incidence of skin cancer is expected to go on rising as a result.

EXTENSION 2
The book *Vital Signs 1996–1997*, compiled by members of the Worldwatch Institute, is a collection of articles and facts about the environment.

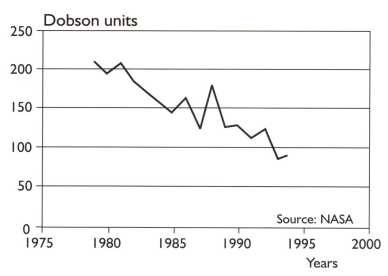

Ozone levels over Antarctica 1979–94

[3] *Vital Signs 1996–1997*, Worldwatch Institute 1996, page 69. Earthscan Publications Ltd, 120 Pentonville Road, London N1 9JN.

3.4 Global warming

The temperature of the earth depends on the balance between the short-wavelength ultra-violet radiation energy from the sun absorbed by the earth, and the re-radiation of long-wavelength infra-red radiation, as shown below.

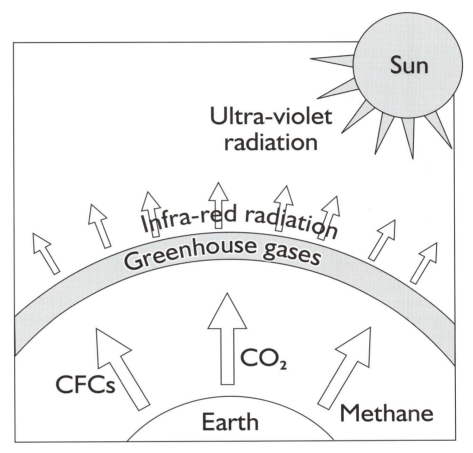

The greenhouse effect

This outward bound infra-red energy is trapped within the atmosphere by water droplets, water vapour and carbon dioxide, and creates a 'greenhouse effect'. The overall effect is that the earth's temperature stays at roughly the same average level – about 15 degrees Celsius (or Centigrade).

Any increase in overall levels of carbon dioxide and other 'greenhouse gases' has the effect of increasing the temperature of the earth, because more infra-red energy is trapped, and less is reflected back. This is global warming.

Many carbon greenhouse gases are produced naturally. However, the activities of humans have increased the amounts of carbon gases collecting in the atmosphere, so tending to upset the natural balance. The main sources of these additional carbon gases are burning forests and scrub-land, power station emissions and vehicle exhausts. The global concentration of carbon dioxide was stable for centuries at around 260 ppm (parts per million), but over the past 100 years it has increased to over 350 ppm. The steady rise since 1959 is illustrated below.[4]

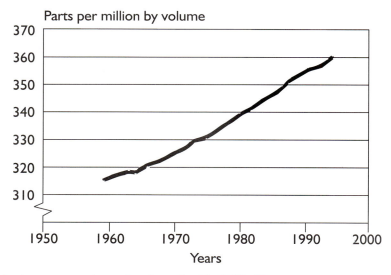

Atmospheric concentrations of carbon dioxide 1959–95

Another important greenhouse gas is methane, which is produced by agriculture (especially rice paddies) and cattle. One molecule of methane has the same greenhouse effect as 30 molecules of carbon dioxide; that is, it traps as much infra-red energy.

The temperature of the earth has stayed fairly constant over the past 10,000 years to within one or two degrees Celsius. Some predictions are that the additional carbon gases will have the effect of increasing the earth's temperature by between three and five degrees over the next fifty years.

[4] Ibid.

Activity 3

3 mins

You may want to consider what effects this could have on normal life. What might be the consequences of a temperature rise of (say) five degrees Celsius? Give it some thought, and jot down your ideas.

_____

_____

_____

_____

_____

_____

_____

_____

As you may have noted, a temperature rise of this order could have devastating effects, including:

- melting glaciers and ice-caps;
- raising sea levels;
- flooding large areas of land in some regions;
- ruining crops through drought in others;
- changing patterns of what would, and would not grow well in particular areas.

The average temperatures recorded from 1950 to 1995, at intervals of five years, are shown below.[5] This does not show a constant rise, but many scientists believe there is a clear warming trend.

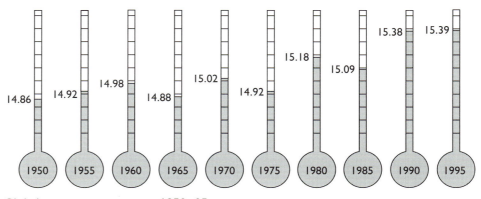

Global average temperature 1950–95

[5] Source of data: ibid.

You may have read reports in the newspapers about global warming, such as the one below.

Global warming theory is now alarming fact, say scientists

GLOBAL warming is no longer theory but a fact; it is largely attributable to man and threatens to worsen significantly in the next century, with serious health implications, a United Nations-sponsored report says.

The Second Assessment Report, presented in Rome yesterday by experts on the Inter-governmental Panel on Climate Change, forecasts a grim future. This is based on findings recorded since a first report was drawn up in 1990 for the Earth Summit in Rio de Janeiro.

Unless we stop burning combustible materials, the experts warn, we can expect average global temperatures to continue to rise at an even greater rate than the half a degree centigrade increase which scientists say they can prove occurred this century.

Over the next 100 years, temperatures are expected to go up by between one and 3.5 degrees – more than the total increase over the past 10,000 years.

Sea levels, which rose by 10–25cm this century, could rise by another four times. The frequency of climatic extremes already experienced this century is likely to increase.

Higher temperatures and humidity would provide fertile conditions for tropical diseases such as malaria and yellow fever. Between 50 and 80 million new cases of malaria are predicted every year over the next century across a vast area, affecting as much as 60 per cent of the world's population.

Climate changes would lead to more deaths from cardio-respiratory disease and heatwaves, scientists say. Specialists from the Universities of Minnesota and Delaware calculate in a study published in Science that a rise of a few degrees would increase the number dying in New York's summer heat from 320 to 880 on average.

Higher sea levels would result in widespread flooding and erosion, affecting tens of millions of people, with many of them forced to migrate.

Moreover, greater hunger and thirst would further aggravate the health of poorer people, who are already plagued by famine and disease.

The conference says the health effects of global warming are not predictions but facts, and that it has already traced recent epidemics to climate change.

In Honduras, malaria-carrying mosquitoes migrated from the south to the north of the country because of desertification caused over the past 20 years, a study published in the *Lancet* was quoted as saying. Cases of malaria in the north went up from 20,000 in 1987 to 90,000 five years later.

Report in *The Daily Telegraph*, 13 December 1995

It should be said that, as with many environmental subjects, there is not a universal consensus about the theories of global warming; some scientists attribute the most recent rises to normal temperature fluctuations.

3.5 Human population growth

The number of human beings occupying the planet Earth now stands at just over 5,700,000,000 (5.7 billion). In 1995, the population increased by some 87 million – nearly one and a half times the current population of the United Kingdom – and is growing by that number every year.

The world population rise over the past 30 years or so is shown below.[6]

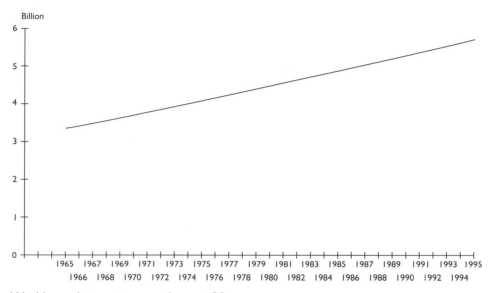

World population rise over the past 30 years

Most of the growth is in the less developed countries. For example, in Africa, the annual birth rate is three times the death rate, whereas in Europe the number of births exceeds the number of deaths by only a small amount each year.

Activity 4

2 mins

What problems is a rising human population likely to bring about, or to make worse? Note down at least one possible problem. Think about the unevenness of the growth in numbers, as well as the overall growth.

_____

_____

_____

[6] Source of data: ibid.

The most obvious problem is that of famine. If the population continues to grow, will food production keep up? In fact, the world is currently producing more than enough food to feed all the humans living in it, at least in terms of calories – the energy provided by food. However, food supplies are not always where they are needed, or, more often, aren't *what* is needed, with cash crops for exports replacing much-needed indigenous food crops. In highly developed countries, food supply generally exceeds demand, but in the poorer areas of the world the reverse is true.

The fact that population growth is greater in the underdeveloped countries tends to make these problems worse. In addition, the increasing number of mouths to feed tends to make the poorer nations even poorer, and there is a growing imbalance of prosperity between the two. Some people talk about a north/south divide. The wealth of the 'north' (North America, Europe, parts of Asia, and including Australasia) contrasts dramatically with the poverty of the 'south' (consisting mainly of South America, Africa, China, and the Indian sub-continent). The north has 25 per cent of the world's people, and 80 per cent of the income; the average life expectancy is 70 years. In the south, 75 per cent of the world population enjoy only 20 per cent of the income, and average life expectancy is about 50 years.

4 Pollution

An **ecosystem** is a biological community of interacting organisms and their physical environment.

Pollution can be described as the contamination of air, water or soil by materials that interfere with human health, the quality of life, or the natural functioning of ecosystems.

Let's look at air pollution first.

4.1 Air pollution

Pollutants of the atmosphere come mainly from industrial plants, power stations and vehicle emissions, although a few are formed naturally.

Activity 5

What unwelcome effects do you think that air pollution is likely to bring? Note down **three** effects.

_____

_____

_____

Air pollutants can cause health and welfare problems for humans, plants and animals. They can also reduce visibility and, as we discussed when looking at acid rain, attack materials such as the surfaces of buildings. Undesirable odours are perhaps among the least harmful effects.

The main air pollutants are as follows.

Pollutant	Comes mainly from
Carbon dioxide	Animals, humans, all combustion sources.
Carbon monoxide	Vehicle exhausts.
Hydrocarbons (including ethane, propane, and ethylene, but not methane)	Vehicle exhausts, solvents, solid waste disposal, industrial processes, fuel combustion.
Lead	Vehicle exhausts, lead smelters, battery plants.
Nitric oxide (NO) Nitrogen dioxide (NO_2) Nitrous oxide (N_2O)	Vehicle exhausts, power stations.
Particulate material	Vehicle exhausts, industrial processes, power stations.
Sulphur dioxide	Power stations burning fossil fuels.

The National Radiological Protection Board, at Chilton, Didcot, OX11 0RQ can give advice about radon.

One natural air pollutant that should be mentioned is radon. This is a colourless, odourless, radioactive gas that is formed naturally in some types of soil, from rock containing uranium. It is linked to lung cancer. People living in some areas in England, including Cornwall, Devon, Somerset, Northamptonshire and Derbyshire, may be liable to exposure from radon.

4.2 Water pollution

Water pollution occurs when water is contaminated by extraneous matter such as:

- infectious agents, such as bacteria and fungi;
- sewage and other waste materials, which decompose and deplete oxygen levels in the water;
- plant nutrients, such as phosphorus, that tend to promote plant growth, so reducing oxygen levels;
- oil, which can enter the sea or other waters from tanker accidents, offshore oil production facilities, seepage, and so on;
- pesticides and other organic chemicals;
- industrial products, such as detergents;
- inorganic minerals and chemical compounds;
- radioactive materials, from nuclear power plants, mining, and the use of radioactive substances in industry, medicine and science.

Measurements of the amount of phosphorus in Lake Windermere showed a steady increase over 30 years. This increase resulted in **eutrophication**: a richness of nutrients supporting a dense plant population, killing animal life by depriving it of oxygen. The process was only reversed when, in 1992, steps were taken to remove phosphorus from the sewage effluent entering the lake.

Activity 6

3 mins

Jot down **two** possible effects of water pollution.

_____

_____

We literally have water on tap in this country, and its quality is high. Should contaminants get into drinking water, they can cause illness or even death. Excessive amounts of nitrates in drinking water, for example, are known to have resulted in the death of infants.

Cholera and typhoid were widespread in Europe a century ago but, largely as a result of improved water supplies and sanitation, they have almost disappeared. In less developed parts of the world, the story is different. In Peru, for example, up to 90 per cent of the population receive a contaminated water supply, and there are hundreds of thousands of cases of cholera each year.

Some pollutants can be absorbed by crops, and so enter the food chain. Metals, including mercury, lead and cadmium, are known to have entered the food chain on occasions, as has arsenic; all of these are highly poisonous if ingested in sufficient amounts.

Another effect of water pollution has already been discussed: acid rain.

Water shortage

Water is frequently in short supply, even though around 1,000 millilitres (40 inches) of rain falls each year in the UK. This is one good reason to cut down on water usage; another is the cost. World-wide, aquifers (layers of rock or soil able to hold or transmit water) are becoming depleted. The demand for water is growing, and is likely to become in seriously short supply in the early 2000s.

5 Chemicals and pollution

EXTENSION 4
The *Consumer's Good Chemical Guide*, by John Emsley, is full of fascinating and useful information.

We cannot avoid chemicals: we use them every day. The human body is itself made of complex chemicals. Chemicals are neither 'good' nor 'bad', but some chemicals are certainly more dangerous to use than others, and chemicals used in the wrong place and in the wrong way can cause untold harm.

More than 68,000 different chemicals are in use in industry and in the home. The majority of man-made chemicals are derived from natural gas or oil.

Activity 7

3 mins

List **three or four** potentially hazardous chemicals in use at your place of work or in your home.

_____

_____

_____

_____

There are so many potentially hazardous chemicals that your list probably won't coincide with mine. Some examples are:

- creosote: a traditional wood preserver, used on fences and so on; it is very poisonous if swallowed or if it comes into contact with the skin;
- paraquat: a weed-killer that is extremely poisonous, and has no known antidote;
- polystyrene: a plastic used for packing and in many other applications. If polystyrene is burned, the fumes can be highly toxic.
- white spirit: used for cleaning and for washing paint-brushes. Some people are sensitive to skin contact with white spirit; it is also flammable. Prolonged exposure to its fumes can cause brain damage.
- asbestos: a fibrous mineral used for insulation, in construction and in car brake linings. There are three forms of asbestos: blue, brown and white. Although all forms of asbestos are very dangerous if inhaled, blue asbestos is the most harmful. Asbestos can cause asbestosis, a serious lung complaint. Because of the hazards, the use of asbestos has decreased in recent years, but there is still a lot of the material left in old buildings.

This list sounds frightening, but of course, for most of the time, and used in the way they were intended, the majority of chemicals used in the home or at work are perfectly safe.

5.1 Accidents involving chemical plants

There have been some devastating accidents involving chemical plants. At Bhopal in India in 1984, 30 tonnes of methyl isocyanate were accidentally leaked from the Union Carbide plant. It was the worst industrial accident in history: over 3,300 people died, 26,000 became chronically ill and the health of an estimated 300,000 to 400,000 other people was affected.

Other major industrial accidents involving chemicals have occurred in Flixborough (England), Seveso (Italy) and in Mexico City in the past few years.

Man-made chemicals seem to be a necessary part of our way of life, but the potential for catastrophe is always there. Even if no one is killed or injured, an industrial accident involving chemicals can have disastrous effects on the environment.

5.2 Agrochemicals

A particular and widespread pollution problem concerns pesticides and other chemicals applied to the land. The world consumption of agrochemicals in 1991 was around 134 billion tonnes. In the UK, something like 4,500 million litres of pesticide are sprayed on crops every year.

World-wide, pests – insects and small animals – spoil up to a third of food crops. Chemical pesticides are therefore used to kill these pests, and can be very effective in doing so. As such they would seem to be a necessary evil.

However, there are two major problems associated with agrochemicals:

■ poisons from pesticides and fertilizers can find their way into water supplies and the surrounding land, causing widespread pollution of the environment, killing fish life and contaminating the food chains of many species;

■ pesticides may kill not only the pests that threaten food crops but the enemies of these pests, and other creatures that eat unwanted weeds.

> In Indonesia in the 1970s, high-yielding strains of rice together with the use of fertilizers and pesticides enabled two rice crops a year to be grown instead of one. However, one result was that the population of a pest called a brown planthopper increased dramatically. To try to get rid of them farmers were spraying each crop up to eight times with pesticide but they still seemed to multiply. Then it was discovered that the pesticides had wiped out all the predators of the brown planthopper, particularly spiders. In other words the pesticide was causing the problem, not helping it.

Now, new biological pest control methods are being developed. These include encouraging the enemies of pests, or, where there are no natural enemies, new species may be introduced into an environment. In the case quoted above, the Indonesian government set up a programme showing farmers how to conserve natural predators of pests, such as spiders. Spraying was stopped, except as a last resort. After three years, huge savings had been made in the cost of pesticides, and rice yields had increased.

Chemicals used as fertilizers have caused problems too. One example is nitrates, used to improve crop yields. Nitrates dissolve easily in water and can get washed from fields into rivers. From there they may contaminate ground water. The EU sets a limit of 50 milligrams per litre for drinking water, and this has sometimes been exceeded.

6 Noise pollution

Almost all activities result in the generation of noise. What do we mean by 'noise pollution', do you think?

Activity 8

3 mins

Write your own definition of 'noise pollution' or an example of it here.

_____

_____

_____

_____

_____

You may have written something like: 'excessive unwanted noise that interferes with normal life'. We all suffer from noise pollution from time to time:

- neighbourhood noise (barking dogs, noisy radios and TVs, slamming doors, loud parties, revving car engines, motor mowers and so on);
- construction noise: pneumatic drills and the like;
- noise at work, from machinery and work activities;
- aircraft noise;
- road traffic noise.

Excessive exposure to loud noise can cause permanent ear damage. This kind of noise is most likely to occur at work. If your ears are ringing when you finish work, or if you have to shout to be heard by someone two metres away from you, then you risk some form of hearing impairment.

The Noise at Work Regulations 1989 are intended to protect people from the risk of hearing damage, by requiring that noise exposure be reduced wherever possible.

EXTENSION 5
A leaflet (NI 207) _Occupational Deafness_ sets out the occupations covered, and other details. It is available from the Department of Social Security.

These Regulations require an employer to make a noise assessment if employees could have a daily exposure of 85 dB (A) or more, and to keep a record. If this is the case, the employees must be told about the dangers from noise exposure, and be offered ear protectors.

If the noise exposures reach 90 dB (A) or more, or 140 dB peak, the noise levels must be reduced where reasonably practicable to do so. These high noise areas must be designated 'ear protection zones' and the employer must ensure that ear protection is worn.

Activity 9

What is the source of the loudest noise at your place of work?

_____

_____

Has any noise assessment been made?   YES   NO

If so, what actions were taken as a result?

_____

_____

_____

If not, and you think an assessment should be made, what action will you take?

_____

_____

_____

_____

People who suffer from substantial permanent hearing loss, as a result of working in certain occupations where noise levels are high, may qualify for industrial injuries disablement benefit.

7 Energy conservation

Energy is the capacity for doing work. Globally, a number of primary sources of energy are employed.

Activity 10

Oil is one primary source of energy. Jot down the names of **three** others.

_____

_____

_____

You may have mentioned:

- coal;
- wood;
- gas;
- wind energy;
- solar energy;
- wave energy;
- hydro-electric energy;
- nuclear energy.

It is inaccurate to say that we are short of energy. The earth receives more energy from the sun in fifteen minutes than humans consume in a whole year.

All energy ultimately comes from the sun; it provides energy both directly and indirectly. Winds, waves, tides and wood all exist because of the sun, and so do fossil fuels like coal and oil that were formed from organic matter millions of years ago.

For most of human history, just about the only extra energy source needed, besides the food we ate to supply bodily energy, was firewood to make camp fires.

When agriculture was developed, and metals were discovered, human needs for energy sources grew considerably. Animals were put into service, water- and wind-driven devices were invented and fossil fuels and coal were burned.

In this century, oil has been a used as an energy source more than any other. More recently, the consumption of natural gas has increased dramatically.

Natural gas, coal and oil are consumed to provide energy – essentially they are burned – and once used they cannot be replaced. These are non-renewable resources.

Anything that can be grown (wood for instance), or sources such as wind and wave power, which don't involve combustion, can be called **renewable energy**.

The figure below shows the percentages of the types of fuel used world-wide.

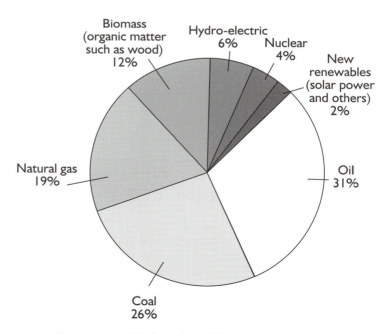

The percentages of the types of fuel used world-wide

Activity 11

We are often reminded to save energy. Give two reasons why efforts should be made to save energy.

_____

_____

You might have mentioned any the following points:

- One problem with fuels that are burned is that they produce carbon dioxide, and this adds to the greenhouse gases.
- Non-renewable energy sources will not last for ever. In the UK, coal stocks may last for another 300–400 years. Over 40 per cent of North Sea oil reserves have already been used. World-wide, oil may last for another 45–75 years at the present rate of consumption.
- The electricity we use is produced almost exclusively by power stations. As mentioned earlier, those that burn coal and oil are the worst offenders when it comes to emitting sulphur dioxide and other substances that cause acid rain and add to the pollution of the atmosphere. (Nuclear power stations do not offend in this way, but many people have misgivings about this method, and are particularly concerned about the dangers of a nuclear accident.)
- Another very practical reason for saving energy is that energy costs money. As we will discuss later in the workbook, a lot of savings can be made by being careful about energy use.

Even with the most modern technology, energy is seldom used very efficiently. Fossil fuel-fired power stations burn fuel to produce steam, which is used to drive turbines; their overall efficiency is less than 40 per cent. The efficiency of a petrol-fuelled car engine is under 20 per cent.

8 The local environment

We have discussed a number of problems in the global environment, but what about the local environment? Most people live, work and play in certain specific areas, and have direct contact with a number of other humans. It is within this environment that they have most influence.

Modern communication systems enable us to expand our horizons considerably: the telephone and, more recently, the Internet, have allowed us to communicate with more people than our ancestors could ever dream of doing. Each of us is able to influence (even if in a very small way!) what goes on in a large portion of the globe. There are many opportunities to make the world a better, or worse, place to live in; the choice is ours.

Activity 12

What aspects of their local environment are important to most people, which make life more, or less, agreeable? Write down at least **three** aspects.

_____

_____

_____

_____

You may have tackled this question by considering the basics first: what do we need to make life tolerable? The following is a suggested starting list of essential human needs:

- wholesome food and drink;
- clothing;
- shelter;
- freedom from fear and oppression;
- to love and be loved.

So far as our surroundings are concerned, most of us seem to be happier where:

- the buildings and other structures in which we work and live are well lit, airy and generally pleasant to look at;
- we have enough space;
- there is freedom from hazards;
- we are not unduly restricted in our movements, or our ability to contact and communicate with others;
- there is opportunity for finding (at least occasional!) periods of peace and quiet;
- temperature and humidity are within certain limits;
- the working or living area is clean, and sanitary facilities are of a high standard;
- we are able to breathe freely, in an unpolluted atmosphere.

As a manager, it will probably be part of your job to ensure that your local working environment is reasonably pleasant for the people working with you. The law makes it clear that every employee is entitled to tolerably comfortable working conditions, as we will discuss in the next session.

Self-assessment I

20 mins

1 'All the problems of this planet are caused by humans.' Say whether you agree with this statement, and explain your reasoning, briefly.

_____

_____

_____

_____

_____

2 Complete the crossword by solving the clues.

Across:

6 Coal, oil and gas. (Two words, 6 and 5)
7 The upper part of the atmosphere, above around 11 kilometres above the earth's surface. (12)
8 Everything around us. (11)
9 A biological community of interacting organisms and their physical environment. (9)
11 Contamination or defilement. (9)
12 The envelope of gases surrounding the earth. (10)
13 Water pollution can occur when bacteria act as an infectious _____. (5)
14 See 11 down.
15 A mountain or hill having openings in the earth's crust through which lava, cinders, steam, gases, etc., are expelled. (7)

Down:

1 A pesticide is usually a _____ and therefore harmful to humans. (6)
2 Ozone acts as a _____ to ultra-violet radiation. (6)
3 The clearance of forests or trees. (13)
4 In the ten years after 1975, the ozone _____ were reported to have fallen by 50 per cent. (6)
5 A dry, barren, often sand-covered area of land. (6)
8 A convulsion of the earth resulting from the release of accumulated stress. (10)
10 Industrial pollutants in precipitation. (Two words 4 and 4)
11 and 14 across. The celestial body on which we live. (Two words 6 and 5)

3   Briefly list the cause and possible consequences of the following environmental concerns. The first one is completed for you.

Name	Cause	Possible consequences
Ozone depletion	Release of CFCs and other man-made gases.	Human health problems, food shortages.
Global warming		
Acid rain		
Continued human population growth		
Continued large-scale use of non-renewable energy sources		
Excessive use of agrochemicals		
Excess noise		

Answers to these questions can be found on pages 84–5.

9 Summary

- We can think of the environment in terms of both what we can influence, and what influences us.

- Natural hazards, such as earthquakes, volcanoes, hurricanes and floods, result in the loss of many lives, and cost a great deal in damage.

- Acid rain is a term used to describe the fall-out of industrial pollutants, mainly sulphur dioxide and nitrogen dioxide, as a result of the burning of fossil fuels such as coal and the smelting of sulphide ores. In the atmosphere they react with water and sunlight, and fall to the earth as dry particles or as acidified rain, snow or fog, sometimes several thousand miles from their source. Acid rain has made lakes unfit for fish, and eats away the surface of buildings.

- The depletion of ozone at the South and North Poles has largely been caused by man-made substances. The result may be increased incidence of skin cancer, eye cataracts and premature ageing. Sources of food could become seriously affected.

- The activities of humans have increased the amounts of carbon gases collecting in the atmosphere, so tending to upset the natural balance between the short-wavelength ultra-violet radiation energy from the sun absorbed by the earth, and the re-radiation of long-wavelength infra-red radiation. This could have devastating effects, including:
  - raising the temperature on the surface of the earth;
  - melting glaciers and ice-caps;
  - raising sea levels;
  - flooding large areas of land in some regions;
  - ruining crops through drought in others.

- The number of human beings occupying the planet Earth now stands at just over 5.7 billion. In 1995, the population increased by some 87 million, and is growing by that number every year.

- Pollutants of the atmosphere come mainly from industrial plants, power stations and vehicle emissions.

- Water pollution occurs when water is contaminated by extraneous matter. Polluted water can cause health problems, either through impure drinking water, or by pollutants being absorbed by crops.

- Chemicals are neither 'good' nor 'bad', but some chemicals are certainly more dangerous to use than others, and chemicals used in the wrong place and in the wrong way can cause untold harm. A particular and widespread pollution problem concerns pesticides and other chemicals applied to the land.

- Noise is a form of pollution. Excessive exposure to loud noise can cause permanent ear damage.

- The local environment is the setting in which each of us has most influence. The achievement of certain standards within this environment can make the difference between misery and potential happiness.

Session B Responding to the problems

1 Introduction

'It is not too fanciful to imagine the next century as one of 'environmental capitalism' in which environmental protection and enhancement is not only a major operational issue and a substantial market, but also a central objective – and source of legitimacy – for both business as a whole and individual enterprises. In this case, the companies taking action now to anticipate environmental opportunities and pre-empt environmental threats may be developing competitive advantage on a timescale measured not in months or years, but also in decades or even centuries.'

Peter James, 'The corporate response'
in M. Charter (ed.) *Greener marketing.*[7]

If you agree that in the twenty-first century environmental protection and enhancement will become a central objective of business, then it would be surely foolish of any organization to ignore the subject now.

We have looked at some major ecological concerns. Now we need to consider the reasons why these things are important to work organizations.

This session is therefore concerned with responses to ecological problems. We will see that:

- governments are responding, by passing laws;
- environmentalists are responding, by being proactive in their approach;
- the public are responding, by demanding that action be taken;
- some work organizations are responding, because they cannot afford to ignore what their customers want.

We begin by taking a look at environmental legislation, and note that industrial companies, in particular, have important legal obligations. Then we examine public opinion as a force for change. Environmentalists, of which we define four types, are discussed next. Lastly in this session, we review the actions that need to be taken to work towards solutions for global ecological problems, and try to decide who is responsible for taking these actions.

[7] Greenleaf, Sheffield, 1992, page 135. Quoted in *Managing the Environment* (Extension 8).

27

2 Environment and the law

Any examination of the environment tends to make one feel somewhat overwhelmed and depressed by the damage potential. However, business organizations have important legal obligations designed to curb the worst excesses.

2.1 Background to the law

Protecting the environment is a global responsibility which needs concerted action between states.

Activity 13

3 mins

Would you expect the United Kingdom Parliament to be free to act independently of other countries, so far as environmental law is concerned?

| YES | NO |

Give a brief reason for your answer.

_____

_____

_____

_____

As we've already discussed, it is not possible for any one country to be self-contained so far as the environment is concerned. It is not surprising, therefore, that the law on the environment is influenced by international bodies. These are:

- the **European Union (EU)**; and to some extent
- the **United Nations**, whose 180 member states pass laws prepared by the **International Law Commission**.

As it has had the more significant impact, we'll focus on the EU.

European legislation is based on the EU's environmental policy, the basic points of which are that:

- **preventative action is to be preferred** to remedial measures;

The message to work organizations is clear: avoid creating pollution – don't think only in terms of clearing up the mess afterwards.

- environmental **damage should be rectified at source**;

  This would mean, for example, that if a river is discovered to contain pollution, the source of that pollution must be found, and steps taken to prevent further contamination.

- the **polluter should pay** for the costs of the measures taken to protect the environment;

  What this means is that organizations causing pollution are held responsible for the costs of preventing or dealing with that pollution. They must recompense anyone who suffers as a result, repair any damage to property, and pay for the costs of putting right the harm to the environment.

- environmental policies should form **a component of the EU's other policies**.

  Most often the EU issues **Directives**; these are then incorporated into UK laws that are passed by Parliament. Some examples of EU Directives are:

- quality standards for air, such as the Smoke and Sulphur Dioxide Directive;
- noise standards, such as the Noise in the Workplace Directive;
- water pollutant emission control, such as the Dangerous Substances in Water Directive.

  Two other important principles are incorporated into UK law. They are:

- **integrated pollution control (IPC)**, which was introduced by the Environmental Protection Act 1990 (EPA);

  This corrects earlier fragmented policies under which different enforcement bodies were responsible for different parts of the environment. Such an unstructured arrangement allowed the possibility that someone who was prevented from (say) dumping waste on land might instead tip it into the sea.

- **BATNEEC – best available technique not entailing excessive cost**.

  It is often possible to reduce the pollution from some process by employing better techniques or equipment. If so, the law says they should be used, unless the polluting organization can demonstrate that the costs of these improved techniques or equipment would outweigh the benefits they would bring. Even then, the polluter might be required to phase in the new equipment over a period of time.

  UK environmental law consists mainly of framework Acts of Parliament, and subsequent Regulations that spell out the detail.

  Let's look at some of the existing legislation related to:

- waste management;
- water pollution;
- air pollution.

29

2.2 The law on waste management

By recycling we usually mean converting waste to reusable material. In Session D, we'll consider what kinds of materials are most suitable for recycling.

Every person and every organization produces waste. Ideally, all waste should be recycled, so that nothing is thrown away. This process occurs naturally: in a forest for example, rotting vegetation and the droppings of animals are 'processed' into nutrients to feed new plants and trees.

Unfortunately, not all waste is recycled naturally. Further, some types of waste are uneconomic to recycle artificially, and still others are so poisonous that they must be disposed of very carefully, to prevent threat to life and damage to ecosystems.

Activity 14

4 mins

How can unwanted waste be disposed of? Suggest **two** means of disposal, and note **at least one** problem associated with each.

_____

_____

_____

_____

_____

Groundwater is rain water that has trickled through rock crevices and pores to form underground pools. Groundwater systems either feed rivers and reservoirs or are tapped directly by water treatment plants.

The main problem with waste is that much of it is active chemically. Some kinds of waste are inflammable, others are poisonous, and some types break down to react with other substances (such as air or water) and produce new toxins.

Most solid waste in the UK is placed in the ground, in so-called **landfill sites**. This method of disposal has at least two main drawbacks:

■ toxic matter can leach into **groundwater** systems, poisoning the water supply;

■ large quantities of methane gas may become trapped and build up, to produce an explosion hazard.

Incineration of waste is expensive, and may pollute the atmosphere. In recent years, use has been made of high temperature incinerators that are able to reduce hazardous wastes to an inert solid residue.

What else can be done with waste? Dump it at sea? Fortunately, this has been banned by international agreement from 1998.

What the law says is that **every organization is responsible for the disposal of its own waste**. The Environmental Protection Act 1990 introduced a new system of waste management, and placed a 'duty of care' on all businesses and other organizations to prevent improper disposal of waste. The following is an extract from a leaflet published by the Department of the Environment:

EXTENSION 6
A DoE Code of
Practice on waste
management is
available, and is listed
in the extension.

When you have waste

When you have waste, you have a duty to stop it escaping. Store it safely and securely.

If you hand waste on to someone else

First, secure it. Most waste should be in a suitable container. Loose material loaded in a vehicle or skip should be covered.

Second, check that the person taking your waste away is legally authorized to do so.

Third, hand over a written description of the waste, and fill in and sign a transfer note for it. The description and note can be the same document.

If you receive waste from someone else

First, you should be legally authorized to accept the waste.

Second, get a written description of the waste from the other person, and complete and sign the transfer note.

Action to take if something goes wrong

You should take action if you suspect that someone else is dealing with waste illegally, before or after it reaches you. Do not give waste to them, or take waste from them, and tell your local council if you are suspicious.

Material with very special problems is **nuclear waste**, from nuclear power stations. Much of this material remains radioactive for thousands of years. The most promising solution seems to be to bury the waste in deep, stable, rock formations, within special containers. However, there is still controversy over the issue.

2.3 The law on water pollution

Any organization wanting to discharge trade or sewage effluent into inland or coastal waters can only do so when given a **consent** by the appropriate Environment Agency. Trade effluents include discharge from industrial plants, farms, and fish farms. A consent also has to be granted before any discharge is made through pipes into the sea.

It is an offence to 'cause or knowingly permit' any such discharge unless it is carried out with a consent.

The 'polluter pays' principle applies, and all costs incurred by the Environment Agency will be recovered from the organization making a discharge.

The new water companies

Until 1990, the supply of water in the UK, and the removal and treatment of sewerage, was effectively under Government control, through the regional water boards. Under the Water Act 1989 these authorities were privatized. There are now 10 water services companies which have responsibility for water supply and sewerage services, plus 29 other smaller ones having responsibility for water supply only. As Simon Ball and Stuart Bell say, in their book *Environmental Law*:

'One stated aim of this (law) was to ... remove the previous cash limits on public spending that had restricted the regional water authorities, thus opening the way for improvements in the quality of water services, but at a cost to the consumer of those services, who will pay for them.'[8]

Perhaps you have views about whether or not public industries should be privatized. The fact remains that, whoever undertakes the task, maintaining a supply of clean water to every home and place of work is an expensive business, as is removing and treating sewage. Either the consumer has to pay for this expense, or the taxpayer does. Pollution control never comes cheap.

2.4   The law on atmospheric pollution

As we have already discussed, pollution of the air can have devastating consequences.

The Environmental Protection Act 1990 introduced a system of Air Pollution Control (APC).

APC is a similar system to IPC (integrated pollution control) but covers only the less polluting substances, and is controlled by local authorities rather than the Environment Agency. It includes such things as:

- lower grade combustion processes;
- small iron and steel furnaces;
- low grade waste incineration;
- animal and vegetable treatment processes.

[8] *Environmental Law*, by Simon Ball and Stuart Bell, 1991, page 289, Blackstone Press Ltd, 9–15 Aldine Street, London W12 8AW.

The following is a brief list of points relevant to air pollution. You might want to make a note in the margin if you suspect that your own organization does not comply with particular aspects.

■ **You must have a licence to operate a potentially polluting industrial process.**

Any organization carrying out an industrial process with the potential to cause pollution, or wanting to get rid of any noxious substance, has to make an application to the Environment Agency for England and Wales, or the Scottish Environmental Protection Agency, or their local authority. The agency will then assess the effects on the environment as a whole, before granting any licence to the applicant.

■ **In a smoke control area, your chimney must not emit any smoke as a result of burning an unauthorized fuel.**

If you live or work in a town or city, your area may have been designated a **smoke control area** under the Clean Air Act 1993. In these areas, it is an offence to burn any unauthorized fuel (such as bituminous coal or wood), and so cause smoke to be emitted from a chimney.

■ **It is an offence for factories and trade premises to emit dark smoke from their chimneys.**

Under the Clean Air Act 1993, dark smoke must not be emitted from chimneys, except when it is unavoidable, such as during start-up. In addition, dark smoke (a shade of grey defined by law) cannot be emitted from fires on industrial or trade premises, or agricultural land.

■ **Air pollution of any kind that is a statutory nuisance may be against the law.**

Under the Environmental Protection Act 1990, a statutory nuisance (in England and Wales) can be:

■ any dust or effluvia arising on any trade or business premises
■ smoke, fume or gases emitted from premises

that are prejudicial to health or a nuisance. The offence also has to be a cause of material harm and to be persistent or likely to recur.

> **Effluvia are unpleasant or noxious odours or exhaled substances affecting the lungs or the sense of smell.**

2.5 Health and safety laws

The subject of the environment is not easily separated from health and safety. After all, many of the effects of the recognized problems of the global environment are to make the world a less safe, and a less healthy, place to live.

Other workbooks in this series deal more specifically with the law on health and safety. Here we have room just to make a quick summary.

Modern health and safety law, like environmental law, is greatly influenced by the EU. And, again in line with environmental law, it consists of framework Acts of Parliament, called 'enabling Acts', and subsequent Regulations that spell out the detail.

The most important framework health and safety Act is the **Health and Safety at Work etc. Act 1974 (HSWA)**. Its purpose is to provide the legislative framework to promote, stimulate and encourage high standards of health and safety at work.

HSWA places obligations on employers to safeguard the health and safety of employees, and of non-employees on their premises. Specifically, employers must ensure that:

- plant and equipment are safely installed, operated and maintained;
- systems of work are checked frequently, so that risks from hazards are minimized;
- the work environment is regularly monitored to ensure that people are protected from any toxic contaminants;
- safety equipment is inspected regularly;
- risks to health from 'natural and artificial substances' are minimized.

In turn, employees are required to act safely, and to have regard to the safety of their colleagues.

EXTENSION 7
If you want to learn more about these Regulations, the approved code of practice: *Workplace health, safety and welfare* is available from HSE Books.

The general guidelines of HSWA are made more specific and focused in the many sets of regulations. A good example to take for our current topic of interest is the **Workplace (Health, Safety And Welfare) Regulations 1992 (WHSWR)**. These apply to a very wide range of workplaces, including schools, hospitals, hotels and places of entertainment, as well as factories, shops and offices. As well as a number of health and safety provisions, they set out certain minimum standards for work conditions, including:

- temperature;
- ventilation;
- lighting;
- room dimensions;
- workstations and seating;
- weather protection;
- toilets;
- washing, eating and changing facilities;
- provision of drinking water;
- clothing storage, and facilities for changing clothing;
- seating;
- rest areas (and arrangements in them for non-smokers);
- rest facilities for pregnant women and nursing mothers;
- maintenance of workplace, equipment and facilities;
- cleanliness;
- removal of waste materials.

This list is referred to in the Work-based assignment, on page 76.

To take but one example, the Approved Code of Practice for the WHSW regulations specifies the minimum number of water closets and washstations in workplaces, according to the number of people working there:

Number of people at work	Number of water closets	Number of washstations
1 to 5	1	1
6 to 25	2	2
26 to 50	3	3
51 to 75	4	4
76 to 100	5	5

Most employers will comply with most provisions of WHSWR without needing to take any action. Nevertheless, through sets of health, safety and environmental regulations like these,

the law is becoming more and more specific about what employers must and must not do.

To avoid prosecution involving fines and possible imprisonment, it pays managers to know the law well.

3 Public opinion

Public awareness of 'green' issues has grown enormously over the past twenty years or so.

Activity 15

3 mins

How does the public express its views? Jot down **three** possible ways in which you might put forward your opinion about an environmental matter.

_____

_____

_____

_____

There are many ways for members of the public to state their views about subjects that concern them. They can:

■ vote in local and national elections for political parties that support their opinions;

So far as environmental issues are concerned, it may be difficult to distinguish between the various parties. Nevertheless, politicians have to be sensitive to the views of the people who vote them to power, and all will generally profess to be 'green' if they perceive that the public want them to be.

■ write to newspapers, radio or television programmes;

Even if they don't, newspaper editors and other journalists always aim to reflect the views of their readers (or listeners, or viewers). If the media do their job well, the ideas they put across will attract the attention of a larger audience, and these views will tend to become more widespread.

■ join lobby groups or protest demonstrations;

People who feel strongly about particular issues, such as:

■ the planned opening of a new road;
■ the announced intention of sinking a drilling rig at sea; or
■ the export of live animals;

may join together to protest. They may do this in a number of ways, such as by lobbying Parliament, or setting up camp on site.

■ respond to public opinion polls, which aim to sample part of the population in order to estimate the numbers in favour of certain views in the whole populace;

If the techniques are correctly applied, and a large enough sample is surveyed, a public opinion poll can be very accurate. However, it is notoriously difficult to frame questions that have the same meaning for everybody.

■ boycott the products and services of companies they don't approve of.

This approach may be the most effective in some cases. It is likely to be applied against organizations that are perceived to be (to give a few examples):

■ polluting the environment;
■ abusing the rights of employees (say) in a third world country;
■ putting profit before environmental concerns;
■ selling environmentally unfriendly goods.

Whatever the means of making their views felt, the public are, directly or indirectly, the customers of nearly every work organization. That being so, it is important to recognize that:

no organization can afford to ignore public opinion, however it is expressed.

36

4 Environmentalists

Biodiversity is simply a contraction of 'biological diversity': the variety of life on earth. In particular, people are worried about a fall in the number of different animal and plant species.

Environmentalists, usually in groups of like-minded people, aim to raise awareness of specific environmental or ecological issues. They are generally concerned about such matters as local or global pollution, and the loss of **biodiversity**. They come in 'all shades of green'.

Typically, environmentalists work by forming pressure groups, political parties, or simply by getting together to protest about a new development that they consider damaging to the environment.

Activity 16

Make a note of **two or three** environmentalist groups that you've heard about, or some of their actions.

_____

_____

_____

_____

_____

_____

Because of the difficulty that minority parties have in becoming elected in the UK, no 'green' party member has yet been elected to Parliament.

You may have mentioned seeing pictures of (or perhaps joining) road protesters at Twyford Down, Hampshire, the M11 extension in north London, the M77 extension in Glasgow, or the Bath and Newbury bypasses.

Or perhaps you recalled the famous incident in 1985, when the Greenpeace ship _Rainbow Warrior_ was bombed and sunk in Auckland Harbour, New Zealand, by French secret service agents. (This event, perhaps more than any other, highlighted the fact that environmentalists frequently upset governments.)

There are four main categories of environmentalists:

- international campaigning organizations, such as *Friends of the Earth*, *World Wide Fund for Nature*, or *Greenpeace*;

- 'green' political parties;

- people who do not choose to join a formal 'green' group, but who feel strongly about environmental issues, and may be influential within their own community or social circle;

- professional scientists, economists, organic farmers and others whose livelihood is directly related to environmentalism.

Real World is a UK group formed in 1996 from more than 30 pressure groups. Its aims, like those of others, can be summarized as a desire to promote **sustainable development**. This is a term that was defined by the World Commission on Environment and Development (WCED) in 1987 as

economic and social development that meets the needs of the current generation without undermining the ability of future generations to meet their own needs.

You will have your own views about environmental activists. Perhaps you think that they do some good, or you may think they are too negative, trying to slow down the progress of civilization without offering any positive alternatives.

There is certainly something to be said for the argument that unjustified and repeated prophecies of 'doom and gloom' may grab the headlines, but do not, in the long term, make people react in a positive way towards the environment. In a recent article (November 1996) under the headline *'Why do we still listen to the doom merchants?'* Matt Ridley made the point that the media give an undue amount of coverage to those who make gloomy predictions about the environment, even when they are repeatedly proved wrong. He says:

'It is odd that people often seem to have more respect for those who fear the worst than for those who state the truth. We are a spectacularly pessimistic species.'[9]

Perhaps you are in sympathy with those who are prepared to be more optimistic about humankind's ability to solve its problems. Or you may agree that it is sometimes very difficult to know where to find the truth about any particular subject in this area.

[9] *The Daily Telegraph* 11.11.96.

5 Improving the environment

From what you have read in this workbook so far, and from your own knowledge and opinions, what practical steps can and should be taken to improve the environment?

Activity 17

4 mins

Make a note of at least three steps of which you would be in favour.

The question in this Activity didn't mention any particular people or organizations, so we don't worry for the moment about who should do the improving.

_____

_____

_____

_____

_____

_____

In general terms, your answer may have been along the lines of the following.

■ To improve the environment, pollution of all media (air, water, and soil), should ideally be eradicated.

This is easier to say than to do. Wherever there are humans, waste products must be disposed of, including:

- agricultural wastes: farm animal manure, and crop residues;
- ashes from the combustion of solid fuels;
- dead animals;
- demolition and construction debris;
- industrial wastes, such as chemicals, paints and sand;
- mining wastes: slag heaps and coal refuse piles;
- nuclear waste;
- sewage-treatment solids.

As we have discussed, some of this waste poses special difficulties, and there are many technological problems to solve.

Nevertheless, much pollution can be reduced. People in the developed world should take the lead, because we have much better opportunities to make ourselves well informed, and have greater resources to take positive actions.

■ To clean up the environment, steps should be taken to reduce or eliminate the emissions that cause acid rain.

In fact, steps have already been taken to do this:

■ an international agreement to reduce sulphur emissions to 70 per cent of 1980 levels by 1993 was achieved in the UK;
■ the rate of nitrogen oxides emissions is frozen at the 1987 level;
■ a reduction in emissions of volatile organic compounds to 70 per cent of 1988 levels by 1999 has been agreed;
■ sulphur emissions are to be reduced in steps to a goal of 20 per cent of 1980 levels by 2010;
■ catalytic converters have been compulsory on all new cars in the United Kingdom since 1993; this helps to reduce the emission of nitrogen oxides.

■ To improve the environment, the use of CFCs should be banned world-wide, to help prevent ozone depletion.

As already remarked, this has been agreed in Europe and the USA for the year 2000.

■ To improve the environment . . . well, what are we going to do about global warming?

Activity 18

3 mins

What practical actions do you believe should be taken to reduce the amount of gases in the atmosphere that result in global warming, such as carbon dioxide and methane? Note down at least one practical step that you think might be carried out.

_____

_____

_____

_____

You may recall the paragraph in the newspaper report we looked at earlier that said:

'Unless we stop burning combustible materials, the experts warn, we can expect average global temperatures to continue to rise at an even greater rate than the half a degree centigrade increase which scientists say they can prove occurred this century.'

It's a lot to expect that everybody will cease burning coal and oil. If these scientists are right, it doesn't seem as if there is much to be done, and that the temperature will inevitably rise. However, we can all do something to reduce **unnecessary** burning of combustible materials. And you may have suggested planting more trees to help reduce the amount of carbon dioxide in the atmosphere.

Every little helps, but it is only at the level of governments that a significant difference can be made. And in a democracy, each of us has a voice in saying what we want our government to do.

■ To improve the **local** environment in which each one of us functions, there's often a great deal that can be done.

We'll go into this subject further in Session D.

6 Responsibilities of organizations and individuals

So, if actions need to be taken, who's responsible?

Activity 19

If you think that 'somebody should do something' about the problems of the environment, who do you think are the people in the best position to take action?

_____

_____

_____

_____

_____

You might have written down any of the following:

■ governments of all countries;
■ governments of the richer nations;
■ environmentalists;
■ large corporations;
■ all companies;
■ all organizations;
■ everyone in the developed world;
■ everyone.

Perhaps all these answers are, to a greater or lesser extent, correct.

Activity 20

3 mins

Here's a slightly different question. Which of the groups in the list above are most likely to be motivated to take action?

	Motivated
■ governments of all countries	☐
■ governments of the richer nations	☐
■ environmentalists	☐
■ large corporations	☐
■ all companies	☐
■ all organizations	☐
■ everyone in the developed world	☐
■ everyone	☐

A saying often quoted by environmentalists is: 'Better to light a single candle than curse the darkness.'

Not everyone, in the developed world or elsewhere, has the motivation to help solve the problems of the environment. Many people are unaware of the issues, and many others are too concerned with their own lives to think about the whole world. There's no immediate reward in (say) disposing of waste carefully, or in planting a tree.

Others feel that there is little impact that one individual can make, as if that excuses their apathy.

Environmentalists are certainly motivated. And democratically elected governments have to respond to public pressure, just as commercial organizations do.

There are increasing pressures and incentives for companies and other organizations to demonstrate that they are concerned about sustainable development, pollution, and other green issues. We will look at what these pressures and incentives are at the start of the next session.

Self-assessment 2

1   Fill in the blanks in the following sentences with suitable words chosen from the list below.

Some significant principles governing EU and UK environmental law are that:

- _____ action is to be preferred to remedial measures;

- environmental _____ should be rectified at source;

- the _____ should pay for the costs of the measures taken to _____ the environment;

- environmental _____ should form a component of the EU's other policies;

- integrated _____ control (IPC), which was introduced by the Environmental _____ Act 1990 (EPA);

- _____ – best available technique not entailing excessive cost.

BATNEEC	DAMAGE	POLICIES
POLLUTER	POLLUTION	PREVENTATIVE
PROTECT	PROTECTION	

2   Indicate whether the following statements are correct or incorrect, and for each incorrect statement, explain briefly why it is wrong.

a   The most common method of solid waste disposal in this country is to place the waste material in landfill sites, because there are no significant hazards associated with this approach.

Correct ☐

Incorrect ☐

The reason this is incorrect is that

_____

_____

_____

b   Small and large companies are all responsible for the disposal of their own waste.

Correct ☐

Incorrect ☐

The reason this is incorrect is that

_____

_____

_____

c It's acceptable to dump toxic materials on waste ground, but you must pay for any damage resulting from your action.

Correct ☐

Incorrect ☐

The reason this is incorrect is that

_____

_____

_____

d The government has taken action to stop organizations from burning fossil fuels by making it against the law to emit smoke.

Correct ☐

Incorrect ☐

The reason this is incorrect is that

_____

_____

_____

e The Health and Safety at Work etc. Act 1974 places duties on employers, not employees.

Correct ☐

Incorrect ☐

The reason this is incorrect is that

_____

_____

_____

f Everyone has a responsibility to help solve the problems of the environment, but not everyone is motivated to do so.

Correct ☐

Incorrect ☐

The reason this is incorrect is that

_____

_____

_____

Answers to these questions can be found on pages 86–7.

7 Summary

- UK law on the environment is influenced by international bodies, notably:

  - the European Union (EU); and to some extent
  - the United Nations.

- Some of the principles behind EU and UK environmental law are that:

  - preventative action is to be preferred to remedial measures;
  - environmental damage should be rectified at source;
  - the polluter should pay for the costs of the measures taken to protect the environment;
  - environmental policies should form a component of the EU's other policies.
  - integrated pollution control (IPC), which was introduced by the Environmental Protection Act 1990 (EPA);
  - BATNEEC – best available technique not entailing excessive cost.

- Three ways of disposing of solid waste are: landfill sites; combustion; and dumping at sea. All have their problems, and some forms of waste, such as nuclear waste, involve solutions that are both very expensive and controversial.

- All organizations are responsible for the safe disposal of their own waste.

- It is an offence to 'cause or knowingly permit' any discharge into water, unless it is carried out with a consent from the Environment Agency.

- There are strict regulations governing the emission of smoke.

- Modern health and safety law, like environmental law, is greatly influenced by the EU. And, again in line with environmental law, it consists of framework Acts of Parliament, and subsequent Regulations that spell out the detail.

- The most important framework health and safety Act is the Health and Safety at Work etc. Act 1974 (HSWA). This enabling Act provides the legislative framework to promote, stimulate and encourage high standards of health and safety at work.

- Through sets of health, safety and environmental regulations, the law is becoming more and more specific about what employers must and must not do.

- There are a number of ways that the opinions of the public can be expressed. Organizations have a vested interest in carefully monitoring public views on environmental matters.

■ A chief concern of environmentalists is sustainable development: 'economic and social development that meets the needs of the current generation without undermining the ability of future generations to meet their own needs'.

■ There are many actions needed to help solve the problems of the environment, and we all have a part to play.

Session C Facing up to the challenge

1 Introduction

EXTENSION 8
This book is listed on page 84.

'... managers must know how to manage their organizations to minimize their detrimental impacts on our environment, and be able to monitor their progress against stated objectives and standards ... Unless managing our environment is viewed as a priority integral to all corporate business strategies, real progress is unlikely.'

John R. Beaumont, Lene M. Pedersen, Brian D. Whitaker, *Managing the Environment*.[10]

Our focus in this session and the next will be on the specific actions that need to be taken by work organizations, in order to meet the challenges posed by the environment.

First, continuing from our discussions in Session B, we try to decide which groups and individuals have some influence over an organization's environmental policies. Then we look at the different possible stances taken by organizations in respect of the environment, and some of the opportunities for businesses.

We then move on to three important topics:

- an environmental policy: what it is and what it should contain;
- what needs to be done to communicate this policy to members of staff;
- the environmental audit, which is a systematic examination of the interaction between any organization and its surroundings.

Quality standards are discussed at length in the workbook *Understanding quality*.

You should note that it is still early days for formal procedures and systems of environmental practice in work organizations to become established. A **British Standard, BS 7750**, has been published, as has the international version **ISO 14001**. As time goes by, these may well become as important to organizations as the quality standards (BS 5750 and BS EN ISO 9000).

---

[10] Butterworth-Heinemann, 1993. Page 42.

47

2 The carrot and stick holders

We ended the previous session with a brief discussion about who was motivated to take action on the environment. We agreed that work organizations are, and now we need to understand what these motivations are.

Public pressure is one factor. But which members of the public are important to organizations?

Activity 21

3 mins

Which of the following groups are important to your organization, and have some influence over its environmental policies?

- customers ☐
- shareholders ☐
- suppliers ☐
- 'green' pressure groups ☐
- employees ☐
- the media (newspaper, television and radio reporters and journalists) ☐
- competitors ☐
- national government ☐
- local government ☐
- the general public ☐

You may have, quite rightly, ticked every box, unless your organization does not have shareholders or competitors.

We have already discussed environmental law; this is one 'stick' by which the **national government** forces organizations to adopt policies that do not result in harm to the environment. **Local government**, along with the Environmental Agency, has a hand in enforcing these laws.

The views of **customers** are obviously very important. As noted earlier, many people will show their disapproval of a company by 'voting with their feet' – that is, going elsewhere for the products and services they want. So this is another 'stick' in one sense, but it can turn into a 'carrot' for those organizations that benefit from such actions. If your organization is seen to be environmentally friendly compared with your rivals, you will obviously have a commercial advantage over them. Thus the actions of **competitors** is also of great consequence.

48

Shareholders – the owners of a limited company, or the authority governing a public company, will usually be most concerned about the organization's financial position. They will not want money spent unnecessarily on environmental policies, but may become very perturbed should those policies result in a down-turn in business.

An organization's **suppliers** of goods and services *may* have some influence, perhaps by indicating that their other customers are setting higher standards.

Green pressure groups and the **media** are most likely to have an impact if the organization acts in a manner that is seen to be damaging to the environment.

What about **employees**? They ought to have a say, because they are the people who have to implement environmental policies. The best way to help people motivated to become whole-hearted about saving energy, to co-operate with recycling schemes, or to avoid waste and pollution, is to get them involved in the decision making.

Lastly, the **general public** are important, because they are potential customers. The organization's 'image' will affect its performance.

3 The challenges and opportunities

So how feasible is it for **you** to be a whole-hearted manager of the environment in your workplace?

Activity 22

3 mins

How does your organization view the environment? Which of the following best sum up the stance it takes? Tick one or more of the boxes, or write down your own summary.

- 'We will seek out the opportunities, and take up the challenges.' ☐

- 'The environment is of no concern or interest to our organization.' ☐

- 'The environment is just another constraint on business: we will do what we have to do, and no more.' ☐

- 'We must continuously review the situation, in case our business is affected.' ☐

- 'The environment could be a financial drain, if we don't watch out.' ☐

- 'Our public image is vital. We must invest to protect and enhance it.' ☐

- 'We must keep up with our competitors.' ☐

- 'This organization takes the view that the environment affects us, and we have an impact on the environment. We must take a responsible and positive attitude.' ☐

49

Your summary of the stance taken by your organization:

_____

_____

_____

_____

_____

_____

We'll discuss environmental policies shortly.

There is obviously a wide range of attitudes and approaches among work organizations. Unless your organization has a declared environmental policy, then it may be difficult for you to gauge just what the common view is.

In a survey of business people, a typical view on the environment was: 'If it's good for my business, I'll do it. Otherwise, they'll have to force me by law to do it.' This is not surprising – after all, the first concern of every business is to make a profit, for lack of profit means failure, and failure inevitably results in extinction. Fortunately, there are many reasons why the environment is good for business.

Activity 23

4 mins

If you work for a business, list **one or more** reasons why widespread concern for the environment could be good for your organization.

_____

_____

_____

_____

_____

You might have made the points that:

- various kinds of 'environmental' activities have become industries in their own right, and industries are capable of making money;

To take a few examples:

- the duty of care on waste means that there are more opportunities for waste disposal companies;
- power generating companies are being required to cut back on the amount of sulphur dioxide they emit through power station chimneys, and so they must pay large sums of money to manufacturers of flue gas desulphurization plants;

- one consequence of the ban on CFCs is that substitutes must be found, manufactured and sold;
- organic food growers are finding business more profitable as people become more conscious of the harmful effects of agrochemicals.

■ as we discussed, a commercial advantage can be gained through appearing 'greener' than your rivals, as customers become more aware of environmental concerns;

■ companies are being forced to 'clean up their act' in terms of waste and pollution; while it may cost money to replace old processes, it does provide an opportunity to increase overall efficiency;

■ as has already been mentioned, money can be saved directly through cutting back on energy; this is also true of water and transport.

4 Environmental policies

EXTENSION 1
This book is listed on page 83.

'For your policy to work, it must be dynamic, flexible, proactive and ongoing.'
— Mark Yoxon, *Successful Environmental Management in a Week*.

The phrase 'environmental policies' has already been mentioned a few times. What do we mean by this?

We can talk about the management of an organization in terms of strategic and tactical management. Broadly speaking, you plan what you want to do, and then find ways of doing it. In order to respond positively to the 'sticks and carrots' of the environment – the challenges and the opportunities to put it another way – a sound **strategy** must be devised.

The expression of such a strategy is the organization's **environmental policy**. This is a statement of its declared approach.

Let's look at one such policy:

In 1992, an eight-point environmental policy was adopted formally by the University of Essex. The eight points are to:

- minimize any disturbance to the local and global environment and to the quality of life of the local community in which the University operates;
- generally the University seeks to be a good neighbour and responsible member of society;
- comply fully with all statutory regulations controlling the University and the sites on which it operates;

- maintain the appearance of the University premises to the highest practical standards;
- take positive steps to conserve resources, particularly those which are scarce or non-renewable;
- assess, in advance where possible, the environmental effects of any significant new development and give due consideration to this in formulating its future plans;
- provide the necessary information to enable staff and students to undertake their work, learning or research properly and with minimal effects on man or the environment;
- keep the public informed of major new developments in the operation and expansion of the University.

Source: Wyvern, University of Essex, 10 June 1992[11]

Activity 24

15 mins

Does your own organization have a declared environmental policy?

YES	NO

Assuming such a written policy exists, get hold of a copy. (If you aren't sure whether your organization has a policy, make some enquiries, perhaps through your manager. In case there is no such document for your organization, it may be that the NEBS Management Centre has their own policy, or can obtain a copy of another organization's document – it's worth asking!)

Read it and compare it with the one for the University of Essex quoted above, and note down **up to three** significant differences between them.

_____

_____

_____

_____

_____

_____

_____

_____

[11] Quoted on page 48 of _Managing the Environment_, by Beaumont, Pedersen and Whitaker, Butterworth-Heinemann, 1993.

An environmental policy should:

- identify which **environmental issues** are important for your organization;
- state, in general terms, what **approach and attitude** you will have towards them;
- acknowledge the **organization's responsibilities** in the local and global community, and say how it will meet these responsibilities;
- summarize the methods by which environmental **performance will be measured** against environmental policies.

Simply writing down a policy is, of course, of little value in itself. The organization also needs to:

- appoint someone, at the most senior level, to take overall responsibility for environmental matters;
- set out environmental objectives, against which performance can be accurately measured;
- review the organization's current and past performance, so that it knows the effects and costs of the actions it has already taken, and can learn from past mistakes;
- be aware of all relevant environmental legislation;
- if considered necessary, arrange for an external audit of the organization's record.

5 Getting the message across

The next important step is to **communicate** the policy and objectives to all staff, and explain what they mean for each individual. If your organization has implemented an environmental policy, perhaps you have the task of ensuring that your team understand its implications, and just what they are expected to do to help turn it into reality.

What do employees need to know?

Activity 25

Let's assume that your employers have recently introduced an environmental policy, appropriate to your organization. What kinds of information would you think you would be expected to pass on to your team?

_____

_____

_____

_____

_____

Perhaps you can answer this question from experience. If so, you may agree that they would probably need:

- one or more sessions in which the policy, and the rationale behind it, is explained at some length;

- specific guidance about how they can help to implement the policy;

  For example, suppose the policy talks of 'conserving resources, particularly those that are scarce or non-renewable'. Team members could join in a discussion about what should be included in that category of resources: fossil fuels and precious metals are among the possibilities here.

- information about the organization's current performance or intended actions;

  An example here could be the information that 'Our heating costs have risen by 15 per cent over the past two years,' or 'We have changed the waste disposal contractors. We expect that the new contractors will charge less for a more efficient service. As part of the arrangement, our waste needs to be segregated as follows …'.

- guidance on how to become more aware, and to monitor their own performance;

  You might perhaps ask for suggestions about how to save energy in the work area. This kind of discussion could be backed up with data on how much has been spent on electricity, gas and oil during the past months or year. A new (realistic) target might be agreed for the coming period.

- instructions regarding systems and procedures.

  An office supervisor might, for instance, give details of a procedure for sending all internal messages by electronic mail, so saving on paper (and helping the environment by causing fewer trees to be cut down).

6 Environmental auditing

Environmental auditing has been defined by the Confederation of British Industry (CBI) as:

' ... the systematic examination of the interaction between any business operation and its surroundings. This includes:

- all emissions to air, land and water;
- legal constraints;
- the effects on the neighbouring community, landscape and ecology; and
- the public's perception of the operating company in the local area.'[12]

[12] Quoted on page 192 of *Managing the Environment*, by Beaumont, Pedersen and Whitaker, Butterworth-Heinemann, 1993.

Although the CBI talks about business operations, the above definition could be applied to any kind of organization.

An environmental audit can be likened to a financial audit: it is a thoroughgoing check or examination of the state of the organization's (financial or environmental) affairs. However, unlike a financial audit, an environmental audit is not compulsory.

Activity 26

10 mins

Thinking about your own place of work, what areas and aspects of the organization's (or department's) activities might be looked into by an environmental auditor? To give you a start, you could perhaps consider:

■ what happens to your company's products after they have been used?
■ what happens to the waste produced in your department or area?

Give this question some thought, and note down your ideas.

_____

_____

_____

_____

_____

_____

_____

_____

_____

_____

Your response to this Activity will depend on several factors, such as: what kind of work is done, what products or services are produced, and how large your organization is.

Among the possible candidates for consideration during an environmental audit are:

■ what you buy, and who you buy it from;
■ the amount of energy and type of energy sources;
■ your waste management;
■ packaging;
■ how your activities affect the local community;
■ your products and services: their environmental impact;
■ aspects of transport.

Let's look at each of these.

- Your suppliers and the materials you buy: are your raw materials obtained from renewable sources and by using environmentally friendly techniques?

  Example: canned tuna fish is a familiar sight at supermarkets. Recently, several of the canning companies have decided to take more care over their sources. Several are now committed to fishing methods that 'protect the marine environment and its species', and they advertise their products as 'dolphin friendly'. In particular, the fishing fleets have to use nets and methods that do not cause dolphins to become trapped and drowned. It is estimated that, between 1959 and 1972, nearly 5 million dolphins died as a direct result of tuna fishing.

- The amount of energy you use, and what sources it is derived from.

  We draw electricity from the national grid, whose inputs are power stations using a variety of fuels, so we can't choose not to use fossil fuels. The only environmentally acceptable approach to energy is to minimize its use, which also happens to be the most cost-effective.

- How you manage your waste products.

  The sort of questions that might be asked here are: 'What happens to solid and liquid waste? Do your work processes pollute the air? What percentage of your waste is recycled?'

- The packaging you use.

  Many suppliers of goods have responded to criticism by reducing the amount of packaging for their products. Others continue to expect their customers to pay for unnecessary packaging, because they know that expensively packaged goods sell better.

- The effects your work activities have on the local community.

Activity 27

3 mins

List **three** ways in which your organization's operations affect the local community.

_____

_____

_____

_____

For most organizations, the effects will be mainly positive. You presumably employ local people, and increase the prosperity of the community. Less desirable effects may include noise; air or water pollution; a contribution to traffic congestion; damage to local roads by heavy vehicles; 'visual' pollution resulting from ugly buildings or other structures and objects in open view.

■ The environmental impact of your products or services.

Example: lead is highly poisonous, especially to the young. This fact has compelled paint manufacturers, and petroleum companies, to reduce the amount of lead in their products, or to eliminate it altogether.

■ The impact of your transport systems on the environment.

A fleet of lorries (for example) can be run efficiently, in which case it will almost certainly use less fuel than one that is not efficient.

Self-assessment 3

15 mins

I Fill in the blanks in the following sentences with words chosen from the list below. (There are more words than blanks.)

a If your organization is seen to be environmentally _____ compared with your rivals, you will have a _____ advantage over them.

b Green _____ groups and the _____ are most likely to have an impact if the organization under scrutiny acts in a manner that is seen to be damaging to the environment.

c The duty of care on waste means that there are more _____ for waste disposal companies.

d An environmental _____ should identify which environmental issues are _____ for your organization, and state, in general terms, what approach and _____ you will have towards them.

e The organization needs to review its current and past _____, so that it knows the effects and costs of the actions already taken, and can learn from past _____ .

ATTITUDE	COMMERCIAL	ENVIRONMENT
FRIENDLY	IMPORTANT	MEDIA
MISTAKES	OPPORTUNITIES	PEOPLE
PERFORMANCE	POLICY	PRESSURE
STRATEGIC		

57

2 Indicate whether these statements are correct or incorrect. Rewrite those that are incorrect so that they read correctly.

a Both national and local government are important to organizations in relation to the environment, for these are the bodies that make the law.

Correct ☐

Incorrect ☐

To be correct, this could be expressed as:

_____

_____

_____

b The widespread concern for the environment could make your organization more profitable.

Correct ☐

Incorrect ☐

To be correct, this could be expressed as:

_____

_____

_____

c An organization's environmental policy is an expression of its declared strategy on the environment.

Correct ☐

Incorrect ☐

To be correct, this could be expressed as:

_____

_____

_____

d Among other things, employees need to know the rationale behind the organization's environmental policy, and guidance on how to measure the performance of their rivals.

Correct ☐

Incorrect ☐

To be correct, this could be expressed as:

_____

_____

_____

e An environmental audit is a systematic examination of the interaction between any organization and its surroundings.

Correct ☐

Incorrect ☐

To be correct, this could be expressed as:

_____

_____

_____

3 Which of the following would be appropriate topics for investigation during an environmental audit?

a Aspects of transport.
b Handling of hazardous materials.
c How you assign work to individuals.
d How your activities affect the local community.
e Packaging.
f Pollution control.
g Physical dimensions of products.
h The amount of energy and type of energy sources.
i The environmental impact of products and services.
j Waste management.
k What you buy, and who you buy it from.
l Your methods of recording purchases.

Answers to these questions can be found on pages 87–8.

7 Summary

- The groups having influence over an organization's environmental policies may include:

  - customers;
  - shareholders;
  - suppliers;
  - green pressure groups;
  - employees;
  - the media;
  - competitors;
  - national and local government;
  - the general public.

- Environmental issues can bring opportunities and challenges. In particular, various sorts of 'environmental' activities have become industries in their own right, in which there are many ways to make money.

- An organization's environmental policy is an expression of its environmental strategy and a statement of its declared approach. It should:

  - identify which environmental issues are important for the organization;
  - state, in general terms, what approach and attitude the organization will have towards them;
  - acknowledge the organization's responsibilities in the local and global community, and say how it will meet these responsibilities;
  - summarize the methods by which environmental performance will be measured against environmental policies.

- An organization's environmental policy and objectives must be communicated to all staff, who need to know, among other things:

  - the rationale behind the policy;
  - how they are expected to help implement the policy;
  - what specific actions are planned;
  - how to monitor their own performance;
  - about systems and procedures.

- Environmental auditing can be defined as the systematic examination of the interaction between any organization and its surroundings. This includes: all emissions to air, land and water; legal constraints; the effects on the neighbouring community, landscape and ecology; and the public's perception of the operating company in the local area. Among the possible candidates for consideration during an environmental audit are:

  - what you buy, and who you buy it from;
  - the amount of energy and type of energy sources;
  - your waste management;
  - packaging;
  - how your activities affect the local community;
  - your products and services: their environmental impact;
  - aspects of transport;
  - pollution control.

Session D The first line manager's role

1 Introduction

'There is a "green" train coming down the corporate track. Some businesses will waste valuable resources, both human and financial, trying to derail this fast-moving force. A few will recognize its importance and growing influence and use it as a vehicle to reach new markets. The train is environmental quality, and I maintain that corporations who climb on board will be the ones that have the best ride . . .'

— Patrick Noonan, 1992, *The Corporate Board*.[13]

Although this green train has been rolling for some time, it has yet to get up to full speed. There is still time for businesses and other work organizations to jump aboard. Those that do not may find themselves at a serious disadvantage, compared with their rivals.

But what can you do, as a first line manager? You may feel you have only limited influence or control over your organization's policies, but there is still a very useful role to play, if you care about the environment.

This session is a discussion of the possible approach you can take to:

- influence the people you work with a view to their giving more consideration to environmental matters;
- undertake constructive activities that will be advantageous to your team and the rest of the organization, while benefiting the environment.

2 What should be done?

There are a number of positive actions that can be taken at a local level, in any workplace, which will promote the environmental cause. In most cases, the organization will also gain from them.

---

[13] Quoted on page 42 of *Managing the Environment*, by Beaumont, Pedersen and Whitaker, Butterworth-Heinemann, 1993.

61

Activity 28

5 mins

Suggest **four or five** areas in which a first line manager should be able to take positive environmental actions.

_____

_____

_____

_____

_____

You could have mentioned some of the points we listed earlier, in regard to environmental auditing (page 54). Among the possibilities are:

- using less energy;
- finding ways to recycle materials, such as metals, plastics and paper;
- making more efficient use of resources, taking into account the potential impact on the environment;
- reducing the amount of waste produced;
- disposing of waste in a less polluting way;
- obtaining materials from renewable sources, or using materials that are less polluting, or that can be recycled after use.

Portfolio of evidence B1.2

Activity 29

15 mins

This Activity may provide the basis of appropriate evidence for your S/NVQ portfolio. If you are intending to take this course of action, it might be better to write your answers on separate sheets of paper.

Thinking back to the subjects we've discussed in the workbook so far, what sort of environmental actions do you think you, as a first line manager, should or could be taking? Answer this question by responding to the following points.

If your organization has a documented environmental policy, briefly describe your role in it.

_____

_____

_____

_____

Whether or not your organization has formally declared its policy on the environment, list **three** practical actions you might take for the work area and team under your control. Do this by ticking one or more of the following boxes, and explaining how you will begin to plan your actions.

■   I plan to reduce the amount of waste material we produce . . .   ☐

     . . . by: _____

_____

_____

■   I plan to dispose of our waste in a less polluting way . . .   ☐

     . . . by: _____

_____

_____

■   I plan to reduce our energy consumption . . .   ☐

     . . . by: _____

_____

_____

■   I plan to change the sources we use for purchases, or use materials that are less polluting, or which can be recycled after use . . .   ☐

     . . . by: _____

_____

_____

■   I plan to make more efficient use of the resources my team use, bearing in mind the impact on the environment . . .   ☐

     . . . by: _____

_____

_____

3 Making savings

Earlier in the unit we discussed reasons for saving energy.

3.1 Saving energy

Remember that:

- every time you turn on a switch the chances are that more fossil fuels will be burned, more greenhouse gases released and that you will make another contribution to the fall-out causing acid rain;
- all energy costs money: wasted energy is wasted money;
- Britain's energy consumption continues to rise, mostly as a result of the activities of industry and commerce.

 Portfolio of evidence B1.2 | **Activity 30** 15 mins

This Activity may provide the basis of appropriate evidence for your S/NVQ portfolio. If you are intending to take this course of action, it might be better to write your answers on separate sheets of paper.

Carry out an energy audit of your area.

- First, find out what forms of energy you consume: oil, gas, electricity, solid fuel and so on.

_____

_____

_____

_____

- Find out if you can, or estimate, the quantities of each fuel type used, and roughly how much each costs.

_____

_____

_____

_____

■ Identify those fuels that are likely to result in most damage to the environment.

_____

_____

_____

■ Ask your team to put forward **three** ideas for cutting down on energy consumption. Estimate or calculate the expected savings for each one. Get them to say exactly how they think each idea might be put into practice.

_____

_____

_____

_____

_____

_____

_____

Many VDU screens are now capable of being switched off automatically when the computers they are attached to have been idle for a period of time.

Some possible ways of saving energy are to:

■ set heating systems to come on only when they are needed;
■ keep doors and windows closed in cold weather;
■ seal doors and windows to prevent draughts;
■ set thermostats to a lower temperature;
■ switch off machines that are not in use;
■ switch off lights when they aren't needed;
■ install insulation in walls and lofts;
■ install double glazing.

3.2 Saving water

Water looks as though it may become a scarce resource, and it certainly will if climate patterns change due to global warming. Water management is very expensive, so the cost to the country as a whole is high. Generally, water supplied to industrial and commercial organizations is metered, so that every drop costs them money.

Activity 31

How might you and your team save water, and so save money? Try to make **one** suggestion.

_____

_____

_____

If you use a lot of water as part of some process, its use will no doubt be monitored. Your expertise may allow you to find ways in which you can save on this commodity, perhaps by re-using it. For other types of organization, it may be a question of getting dripping taps repaired, or cutting down on the use of water in (for example) gardens.

4 Recycling

There are two main reasons why recycling is carried out. One is financial; for example, people in the developing world generally produce less waste, and recycle more of it, through economic necessity. The second reason is a genuine desire to help the environment, which is more likely to be encountered in the developed world.

In the UK, around 25 per cent of industrial waste is recycled, and only 5 per cent of domestic refuse. The materials from people's homes that are most cost-effective to recycle are **glass**, **paper**, and **aluminium cans**.

It is perfectly possible to recycle many types of **plastics**, but consumers generally find it difficult to separate one type from another. Manufacturers and commercial users are usually in a far better position to collect plastic materials for recycling.

Recycling may be carried out by the organization producing the waste, or by companies that specialize in recovering materials.

Waste can even be burned to good purpose. Electricity can be generated using power stations that are able to burn:

- rubbish;
- methane gas produced by decomposing waste;
- waste matter from agricultural processes, called biofuels;
- coppiced wood from forests grown for this purpose.

Among other materials, gold, silver, tin, copper, zinc, solder alloy and cadmium can all be recovered from waste products using electrolysis.

How does your organization fare with recycling?

Activity 32

10
mins

This Activity may provide the basis of appropriate evidence for your S/NVQ portfolio. If you are intending to take this course of action, it might be better to write your answers on separate sheets of paper.

Which materials does your organization recycle, or collect for recycling?

_____

_____

_____

_____

_____

Which materials that your team uses might be recycled, but currently aren't?

_____

_____

_____

_____

_____

How could you find out more about the possibility of recycling more of your waste products?

_____

_____

_____

_____

_____

How will you go about implementing an updated recycling scheme? Think about how to make sure that your proposal is seen or heard by the appropriate person, and write down how you will proceed.

_____

_____

_____

_____

_____

5 A healthy and safe workplace

This is one area where the standards you set as an individual manager can really make a difference.

5.1 Cleaning up the work area

No matter how stringent the requirements for cleanliness are in your work area, the chances are that they can be improved.

Activity 33

List **three** benefits of having clean conditions for people to work in. Answer in terms of your own workplace.

_____

_____

_____

What could be done to clean up your own work area? Try to suggest **three** actions that might be taken.

_____

_____

_____

Working in clean conditions often brings benefits of:

- increased hygiene, making it a healthier place to work;
- less clutter, reducing the risk of accidents;
- improved morale;
- greater efficiency: a clean uncluttered area can help create a sense of pride in the work.

In some jobs, cleanliness is essential, and cleaning up has to become part of the normal routine. Some obvious examples are hospitals, doctors' surgeries and so on, and places where food is served.

In many other kinds of workplaces, it isn't unusual to find a less than rigorous attitude and approach to cleanliness. Conditions range from general untidiness to the long-term accumulation of filth. If you can't persuade your team to get into the habit of cleaning up at the end of a day or a task, you may need to organize periodic clean-up campaigns.

5.2 Making a safer workplace

As we discussed in Session B, stringent laws make it compulsory for employers to safeguard the health and safety of their employees. We looked briefly at the Health and Safety at Work etc. Act 1974, and the Workplace (Health, Safety and Welfare) Regulations 1992. There are a number of other health and safety laws that are also relevant to care of the environment, including the:

- Control of Asbestos at Work Regulations 1987;
- Control of Lead at Work Regulations 1980;
- Control of Pesticides Regulations 1987;
- Control of Substances Hazardous to Health (COSHH 2) Regulations 1994;
- Noise at Work Regulations 1989.

Complying with the law is important and necessary, but is it enough? Even more crucial, perhaps, is an attitude of mind that encourages people always to be on the lookout for ways of making the workplace safer. There's always plenty that can be done: vigilance is the watchword.

Activity 34

4 mins

Try to suggest **one** action that you could take, or recommendation that you might put forward, to help make the area under your control safer for the people who work there. Relate your suggestions to the environment.

_____

_____

_____

_____

_____

According to the law, health and safety training should: be repeated periodically where appropriate; be adapted to take account of new or changed risks; take place during working hours.

Your ideas about safety will be relevant to the kind of work you do, and the particular circumstances and conditions of your job. Among the many possible responses to this Activity are:

- taking greater precautions with handling any hazardous substances that you use, store or transport;
- improving work conditions, so that people are under less stress, and therefore less prone to accidents;
- enforcing rules about tidiness, to help reduce the risk of tripping accidents;
- providing more training, so that team members have a better understanding of how to work safely, while helping to protect the environment;
- implementing systems for monitoring the levels of hygiene.

6 Setting the right example

Having nearly reached the end of this workbook, you may (or may not) be more convinced about the need to treat the world in which we live with greater respect. It can sometimes be difficult to get to the truth about particular subjects, especially when even the experts disagree (Is global warming an inevitability? Will food production keep up with the rising population?) But some facts are indisputable. These include the reality that:

- we would be foolish to think that 'the plague of humans' settled on this planet can continue to exploit it indefinitely and indiscriminately;
- what is good for the environment is, by and large, good for us;
- even if we ignore the global environment, all of us can make our local environment a better place for ourselves and our fellows in a number of ways, including cutting back on waste and pollution;
- environmental law places strict controls on work organizations;
- organizations that try to fudge their approach to the environmental cause can expect to get a bad press, and will probably lose business as a result;
- there are commercial benefits available in the 'the environmental industry'.

For these reasons at least, our environmental responsibilities are inescapable. That being so, you as a first line manager have an important role to play.

Perhaps the most crucial aspect of this role is leadership by example. It will no doubt be up to you to communicate the organization's policies and programmes to your team members, and to show that you are committed to the environmental cause.

Self-assessment 4

15 mins

1   Briefly explain what recycling means, and give an example of the way in which it can be beneficial to the environment.

_____

_____

_____

_____

2   List three kinds of savings that an organization might make, that are consistent with a positive and helpful attitude to the environment.

_____

_____

_____

3 Briefly explain why you do, or do not, agree with the following statements, by giving an example.

a Health and safety, and the environment, are closely linked.

_____

_____

_____

b All health and safety laws are also relevant to environmental laws.

_____

_____

_____

c All materials can be recycled back to their original state.

_____

_____

_____

d A clean and tidy workplace is a healthier workplace.

_____

_____

_____

Answers to these questions can be found on page 88.

7 Summary

- Among the possible actions that might be taken by a first line manager to help protect and enhance the environment are the following:

  - use less energy, or obtain energy from renewable sources;
  - find ways to recycle materials, such as metals, plastics and paper;
  - make more efficient use of resources, taking into account the potential impact on the environment;
  - reduce the amount of waste produced;
  - dispose of waste in a less polluting way;
  - obtain materials from renewable sources, or use materials that are less polluting, or that can be recycled after use.

- Some possible ways of saving energy are to:

  - set heating systems to come on only when they are needed;
  - keep doors and windows closed in cold weather;
  - seal doors and windows to prevent draughts;
  - set thermostats to a lower temperature;
  - switch off machines that are not in use;
  - switch off lights when they aren't needed;
  - install insulation in walls and lofts;
  - install double glazing.

- Saving water can be beneficial to the environment, and will reduce the organization's costs.

- Recycling may be carried out within the organization that produced it, or by a specialist firm. All kinds of materials can be recycled, including rubbish.

- Working in clean conditions often brings benefits of:

  - increased hygiene, making it a healthier place to work;
  - less clutter, reducing the risk of accidents;
  - improved morale;
  - greater efficiency: a clean uncluttered area can help create a sense of pride in the work.

- A safe and healthy workplace is usually consistent with an environmentally friendly workplace.

Performance checks

Jot down the answers to the following questions on *Managing a Safe Environment*.

Question 1   Acid rain can be called an ecological problem. Why is the word 'ecological' used here, rather than simply 'environmental'?

_____

_____

Question 2   In a sentence or two, explain the phenomenon of ozone depletion.

_____

_____

Question 3   Explain what you understand by 'pollution'.

_____

_____

Question 4   What are the two main problems associated with agrochemicals?

_____

_____

Question 5   Name two fossil fuels, and two sources of renewable energy.

_____

_____

Question 6   Which body outside of this country has most influence in our environmental law-making?

_____

_____

Question 7   Where does most air pollution come from?

_____

_____

Question 8   What, in brief, are the implications for organizations of the duty of care on waste?

_____

_____

Question 9   The Health and Safety at Work etc. Act 1974 (HSWA) is an 'enabling' Act. What does that mean?

_____

_____

Question 10   Briefly explain why public opinion is so important to organizations.

_____

_____

Question 11   What do we mean by 'sustainable development'?

_____

_____

Question 12   If someone asks you what should be contained in an organization's environmental policy document, what will you tell them? Write down two points.

_____

_____

Question 13   How would you define environmental auditing?

_____

_____

Question 14   List three possible ways to save energy in the workplace.

_____

_____

Question 15   Suggest one way in which ordinary domestic waste might be recycled.

_____

_____

Answers to these questions can be found on pages 89–90.

2 Workbook assessment

60 mins

Read the following case incident and then deal with the instruction that follows, writing your answers on a separate sheet of paper.

■ Driftwood Manor is a residential sixth form college, catering mainly for 16–18 year-olds planning to go on to University. It offers courses in music, theatre, art and sport, as well as a number of academic subjects. The college is situated in a large house on the outskirts of a provincial town, with several acres of grounds. It has a high reputation, both for its record of A-level passes, and the considerable range of extramural activities, including riding, shooting and cross-country running.

The Principal, Doctor Steven Cogwood, is concerned about global ecological problems, and encourages his students to take an interest in these matters. He is keen to embark on a structured plan and a programme of action, designed to enable the college to play its full part in protecting and enhancing the environment.

Assume that you, as an expert who happens to take a great interest in the school, have agreed to draw up an outline plan to be considered at the next Governors' meeting. You will need to:

■ draft a suitable environmental policy document;
■ put forward three practical suggestions for environmentally friendly actions that the college could take;
■ suggest ways in which the students themselves can be encouraged to become involved.

As you have limited knowledge of the college and its activities, make any assumptions you wish, but be sure to state what these assumptions are.

3 Work-based assignment

The time guide for this assignment gives you an approximate idea of how long it is likely to take you to write up your findings. You will find you need to spend some additional time gathering information, perhaps talking to colleagues and thinking about the assignment.

Your written response to this assignment should form useful evidence for your S/NVQ portfolio. The assignment is designed to help you to demonstrate your Personal Competence in:

- building teams;
- focusing on results;
- thinking and taking decisions;
- striving for excellence.

On page 34, we discussed the fact that the Workplace (Health, Safety And Welfare) Regulations 1992 (WHSWR) set out certain minimum standards for work conditions. The list was:

- temperature;
- ventilation;
- lighting;
- room dimensions;
- workstations and seating;
- weather protection;
- toilets;
- washing, eating and changing facilities;
- provision of drinking water;
- clothing storage, and facilities for changing clothing;
- seating;
- rest areas (and arrangements in them for non-smokers);
- rest facilities for pregnant women and nursing mothers;
- maintenance of workplace, equipment and facilities;
- cleanliness;
- removal of waste materials.

**What you have
to do**

Choose five aspects of work conditions from this list, and, for each one, find out how well your work area complies with the law. Involve your team in this exercise.

To do this, you will either have to get help from someone in your organization who has knowledge of the law, or else obtain a copy of *Workplace health, safety and welfare*: the Approved Code of Practice for the Workplace (Health, Safety and Welfare) Regulations 1992, and study it. This document (listed as an extension on page 83) is available by mail order from HSE Books; telephone number 01787 881165; fax 01787 313995.

Write a short report on your findings, together with any recommendations you may have, and address it to your manager.

Reflect and review

1 Reflect and review

Now that you have completed your work on *Managing a Safe Environment*, let us review our workbook objectives.

The first objective was:

- When you have completed this workbook you will be better able to identify many of the world's ecological problems, and recognize their causes.

In Session A, we looked at a number of global ecological problems: acid rain; global warming; ozone depletion; and human population growth. In addition, we reviewed the types of pollution (air, water and land pollution).

It is obviously important that action should be taken to stop polluting the environment, but there are few easy completely 'clean' methods of recycling with waste that are also inexpensive. Some manufactured chemical compounds in particular are difficult to deal with, and nuclear waste is radioactive for thousands of years.

There are many sources of energy, but the burning of fossil fuels increases the levels of carbon dioxide in the atmosphere, so reducing energy consumption also helps the environment. Noise is a pollutant of a different kind, and is sometimes a health hazard in work places.

Look at the list of extensions on pages 83–4.

- Do you feel you are sufficiently well informed about these topics? If not, how will you set about finding out more?

_____

_____

- How could you increase your team's awareness of the issues?

_____

_____

The second objective was:

■ When you have completed this workbook you will be better able to summarize the response to these problems by governments, environmentalists, the general public, the media, and other groups.

We briefly discussed the principles of current environmental law, and the fact that much of the impetus in this area is the result of the UK's membership of the EU.

Public opinion is a force that drives many engines; in particular it affects governments, through the election of environmentally aware politicians; the media, who are quick to react to and reflect the opinions and beliefs of their viewers, readers and listeners; commercial organizations, who must respond to the wants of their customers.

Environmentalists are the most active group of people on the scene, and some high-profile activities have raised public awareness.

> The Work-based assignment on page 76 asks you to look into one aspect of environmental law.

■ **Are you confident that the work area and activities under your control comply with environmental law? Should you consult your manager about this?**

_____

_____

The third objective was:

■ When you have completed this workbook you will be better able to explain why the environment is important to work organizations, and say what kind of actions are appropriate for them to take.

As the author of the quotation in our workbook introduction stated, the environment 'is a new fact of business life'. It can no longer be ignored: all organizations are being compelled, through legislation and the force of public opinion, to play their part in protecting the environment.

We can talk about 'environmental problems' or 'the environmental challenge': all difficulties bring opportunities, and there are many ways in which organizations can gain from environmental issues.

We discussed the setting out of an environmental policy and objectives, and implementing them through the involvement of all employees.

■ **Is your organization making plans and taking action? How might you use your influence to persuade your colleagues to do more?**

_____

_____

The last objective was:

■ When you have completed this workbook you will be better able to take a number of positive steps in your own work area to protect and enhance the local and global environment.

We discussed several possible steps that might be taken by a first line manager, including: finding ways to make savings on energy, and water; recycling more materials; cutting down on waste, and dealing with it more effectively; keeping the work area clean and tidy; and setting the right example.

■ **What else can you do, in your job, to further the environmental cause?**

_____

_____

■ **What further actions can you take to involve your team and motivate them to do more?**

_____

_____

2 Action plan

Use this plan to further develop for yourself a course of action you want to take. Make a note in the left-hand column of the issues or problems you want to tackle, and then decide what you intend to do, and make a note in Column 2.

The resources you need might include time, materials, information or money. You may need to negotiate for some of them, but they could be something easily acquired, like half an hour of somebody's time, or a chapter of a book. Put whatever you need in Column 3. No plan means anything without a timescale, so put a realistic target completion date in Column 4.

Finally, describe the outcome you want to achieve as a result of this plan, whether it is for your own benefit or advancement, or a more efficient way of doing things.

Desired outcomes					Actual outcomes
1 Issues	2 Action	3 Resources	4 Target completion		

3 Extensions

Extension 1

Book   *Successful Environmental Management in a Week*
Author   Mark Yoxon
Edition   1996
Publisher Headway – Hodder and Stoughton

The introduction says: 'This practical book has been written for managers who want to cut through the green rhetoric and get to the heart of practical environmental management action that will make sense for their business.' It is small, slim, and well-written.

Extension 2

Book   *Vital Signs 1996–97*
Edition   1996
Publisher The Worldwatch Institute

This is a book of up-to-date facts and comment, covering all aspects of the environment. It is published annually, so you should try to get hold of the latest edition.

Extension 3

CD-ROM   *Encarta 97 Encyclopaedia* – World English Edition (*Encarta* is published annually – get the latest edition you can.)
Publisher Microsoft Corporation

One word of warning – while investigating environmental topics, you may find yourself getting enjoyably distracted by many of the thousands of articles covering quite irrelevant subjects.

Extension 4

Book   *The Consumer's Good Chemical Guide*
Author   John Emsley
Edition   1996
Publisher Corgi Books

Winner of the Rhône-Poulenc Science Book Prize, this book has been described as 'A book about chemistry that is fun to read'. It puts right several misconceptions about which chemicals are, and are not, polluting.

Extension 5

Leaflet   (NI 207) *Occupational Deafness*
Publisher Department of Social Security

Extension 6

Title   *Code of Practice on Waste Management*
Publisher Department of the Environment

Extension 7

Title   *Workplace health, safety and welfare – Approved Code of Practice*
Publisher HSE Books

Extension 8

Book   *Managing the Environment*
Authors   John R. Beaumont, Lene M. Pedersen, and Brian D. Whitaker
Edition   1993
Publisher Butterworth-Heinemann

Though interesting, in that it is written from the manager's viewpoint, you may find this book rather heavy going as a background read to the subject.

These Extensions can be taken up via your NEBS Management Centre. They will either have them or will arrange that you have access to them. However, it may be more convenient to check out the materials with your personnel or training people at work – they may well give you access. There are other good reasons for approaching your own people; for example, they will become aware of your interest and you can involve them in your development.

4 Answers to self-assessment questions

Self-assessment 1 on page 23

1   This statement can be interpreted in more than one way; in particular, it depends what is meant by a 'problem'. However, it must be acknowledged that not all disasters are caused by humans – volcanoes are not, and neither are earthquakes, for example. (It has been said that planet Earth is suffering from a 'plague of humans', but that given time it will recover.)

2   The completed crossword is as follows.

3   Some possible answers are shown below.

Name	Cause	Possible consequences
Ozone depletion	Release of CFCs and other man-made gases.	Human health problems, food shortages.
Global warming	Mostly excess carbon dioxide, probably the result of industrial processes, causing heat to be 'trapped' in stratosphere.	Raising of sea levels and consequent flooding; drought in other areas.
Acid rain	The fall-out of industrial pollutants such as sulphur dioxide and nitrogen dioxide, combining with water in the atmosphere.	Making lakes unfit for fish; damage to building surfaces; possible destruction of trees.
Continued human population growth	(We didn't discuss a 'cause', and this is a big and difficult subject. Some would say that increasing women's access to health care, education, and income-generating employment is the only sure way to slow population growth.)	An uneven reduction in world-wide living standards.
Continued large-scale use of non-renewable energy sources	The 'cause' here is the insatiable demand for more energy.	The ultimate collapse of civilization, unless alternative sources are developed.
Excessive use of agrochemicals	The desperation of poor farmers? Lack of government control? Pressures to increase yields?	Pollution of water supplies; disruption of the environment of animal and insect life.
Excess noise	Proximity to machinery.	Hearing damage.

Self-assessment 2 on page 43

1 The completed sentences are:

- PREVENTATIVE action is to be preferred to remedial measures;
- environmental DAMAGE should be rectified at source;
- the POLLUTER should pay for the costs of the measures taken to PROTECT the environment;
- environmental POLICIES should form a component of the EU's other policies;
- integrated POLLUTION control (IPC), which was introduced by the Environmental PROTECTION Act 1990 (EPA);
- BATNEEC – best available technique not entailing excessive cost.

2 The correct statements and corrected wrong statements are as shown.

a The most common method of solid waste disposal in this country is to place the waste material in landfill sites, because there are no significant hazards associated with this approach.

Correct ☐
Incorrect ☒

The reason this is incorrect is that

although the first part of the sentence is correct, there are several hazards associated with landfill site waste disposal. They include water pollution and methane build-up.

b Small and large companies are all responsible for the disposal of their own waste.

Correct ☑
Incorrect ☐

c It's acceptable to dump toxic materials on waste ground, but you must pay for any damage resulting from your action.

Correct ☐
Incorrect ☒

The reason this is incorrect is that

it is never acceptable to dump toxic materials, whether or not the polluter pays.

d The government has taken action to stop organizations from burning fossil fuels by making it against the law to emit smoke.

Correct ☐
Incorrect ☒

The reason this is incorrect is that

although it is illegal in most circumstances to emit dark smoke, light-coloured smoke is permitted; in addition, many fossil fuels are 'clean' in the sense that they do not emit smoke.

e The Health and Safety at Work etc. Act 1974 places duties on employers, not employees.

Correct ☐

Incorrect ☒

The reason this is incorrect is that

HSWA places duties on both employees and employers.

f Everyone has a responsibility to help solve the problems of the environment, but not everyone is motivated to do so.

Correct ☑

Incorrect ☐

Self-assessment 3 on page 57

1 The completed sentences are:

a If your organization is seen to be environmentally FRIENDLY compared with your rivals, you will have a COMMERCIAL advantage over them.

b Green PRESSURE groups and the MEDIA are most likely to have an impact if the organization under scrutiny acts in a manner that is seen to be damaging to the environment.

c The duty of care on waste means that there are more OPPORTUNITIES for waste disposal companies.

d An environmental POLICY should identify which environmental issues are IMPORTANT for your organization, and state, in general terms, what approach and ATTITUDE you will have towards them.

e The organization needs to review its current and past PERFORMANCE, so that it knows the effects and costs of the actions already taken, and can learn from past MISTAKES.

2 a Both national and local government are important to organizations in relation to the environment, for these are the bodies that make the law.

Correct ☐

Incorrect ☑

To be correct, this could be expressed as:

Both national and local government are important to organizations in relation to the environment: national government makes the laws, and local government help to uphold them.

b The widespread concern for the environment could make your organization more profitable.

Correct ☑

Incorrect ☐

c An organization's environmental policy is an expression of its declared strategy on the environment.

Correct ☑

Incorrect ☐

87

    d Among other things, employees need to know the rationale behind the organization's environmental policy, and guidance on how to measure the performance of their rivals.

Correct ☐   Incorrect ☑

To be correct, this could be expressed as:

Among other things, employees need to know the rationale behind the organization's environmental policy, and guidance on how to measure their own performance.

    e An environmental audit is a systematic examination of the interaction between any organization and its surroundings.

Correct ☑   Incorrect ☐

3   The following would be appropriate topics for investigation during an environmental audit,

    a Aspects of transport.
    b Handling of hazardous materials.
    d How your activities affect the local community.
    e Packaging.
    f Pollution control.
    h The amount of energy and type of energy sources.
    i The environmental impact of products and services.
    j Waste management.
    k What you buy, and who you buy it from.

Self-assessment 4 on page 70

1   Recycling means converting waste to reusable material. An example (out of many possible ones) is the re-processing of waste paper, which results in fewer trees being cut down.

2   You could have mentioned saving on energy, water and transport.

3   a Health and safety, and the environment, are closely linked.

This is certainly true; an example is the fact that many chemicals that are hazardous to humans will also damage the environment.

    b All health and safety laws are also relevant to environmental laws.

This is not true. Many health and safety laws are relevant to environmental laws, but some are not; an example is the law on the manual handling of goods.

    c All materials can be recycled back to their original state.

This is not true; for example, old wood can't be recycled back to trees.

    d A clean and tidy workplace is a healthier workplace.

True: there is less chance of bacteria collecting in a clean work area, for example.

5 Answers to the quick quiz

Answer 1 Ecology is the study of the interaction of people with their physical environment. So the word 'ecological' is appropriate and more accurate, because acid rain (like many other problems) is the result of human interaction with the environment.

Answer 2 Ozone, a modified form of oxygen, acts as a filter to ultra-violet radiation from the sun. There is now a depletion of ozone, largely caused by man-made substances, especially chemicals called chlorofluorocarbons (CFCs).

Answer 3 Pollution can be described as the contamination of air, water or soil by materials that interfere with human health, the quality of life, or the natural functioning of ecosystems.

Answer 4 The two major problems associated with agrochemicals are that:

- poisons from pesticides and fertilizers are liable to find their way into water supplies and to other areas, causing widespread pollution;
- pesticides may kill not only the pests that threaten food crops but the enemies of these pests, and other creatures that eat unwanted weeds.

Answer 5 You could have mentioned coal, oil and gas as fossil fuels. Renewable energy sources include wind power, wave power and hydro-electric power (among others).

Answer 6 The European Union has most influence over our environmental legislation.

Answer 7 Pollutants of the atmosphere come mainly from industrial plants, power stations and vehicle emissions.

Answer 8 The duty of care requires that organizations take responsibility for their own waste, and must dispose of it properly, following a laid down procedure.

Answer 9 The purpose of an enabling Act is to provide a framework, upon which may be added Regulations that spell out the detail.

Answer 10 For most organizations, the public are customers or potential customers. If their public image is lowered – say, by being seen as polluters of the environment –businesses will find it harder to sell their products and services. Other organizations may find life more difficult in other ways. For example, the managers of a hospital would be under great pressure to respond to public criticism.

Answer 11 Sustainable development has been described as: 'economic and social development that meets the needs of the current generation without undermining the ability of future generations to meet their own needs'.

Answer 12   An environmental policy should: (a) identify the environmental issues that the organization considers important; (b) state, in general terms, the approach and attitude it will have towards them; (c) acknowledge the organization's responsibilities in the local and global community, and say how it will meet these responsibilities; (d) summarize the methods by which its environmental performance will be measured against environmental policies.

Answer 13   According to the CBI, environmental auditing is 'the systematic examination of the interaction between any business operation and its surroundings. This includes: all emissions to air, land and water; legal constraints; the effects on the neighbouring community, landscape and ecology; and the public's perception of the operating company in the local area.

Answer 14   As we discussed, you might: set heating systems to come on only when they are needed; keep doors and windows closed in cold weather; seal doors and windows to prevent draughts; set thermostats to a lower temperature; switch off machines that are not in use; switch off lights when they aren't needed; install insulation in walls and lofts; install double glazing. You may have other good ideas for saving energy.

Answer 15   It can either be burned, or put into methane-producing plants, and so turned into electricity.

6 Certificate

Completion of this certificate by an authorized person shows that you have worked through all the parts of this workbook and satisfactorily completed the assessments. The certificate provides a record of what you have done that may be used for exemptions or as evidence of prior learning against other nationally certificated qualifications.

Pergamon Open Learning and NEBS Management are always keen to refine and improve their products. One of the key sources of information to help this process are people who have just used the product. If you have any information or views, good or bad, please pass these on.

N E B S

M A N A G E M E N T

D E V E L O P M E N T

SUPER **S E R I E S**

T H I R D   E D I T I O N

Managing a Safe Environment

....

has satisfactorily completed this workbook

Name of signatory   ....

Position   ....

Signature   ....

Date   ....

Official stamp

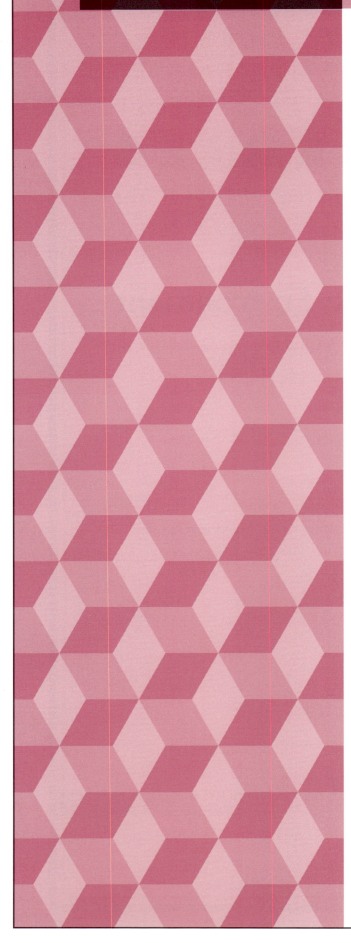

SUPER SERIES

SUPER SERIES 3

0-7506-3362-X   Full Set of Workbooks, User Guide and Support Guide

A. Managing Activities

0-7506-3295-X	1. Planning and Controlling Work
0-7506-3296-8	2. Understanding Quality
0-7506-3297-6	3. Achieving Quality
0-7506-3298-4	4. Caring for the Customer
0-7506-3299-2	5. Marketing and Selling
0-7506-3300-X	6. Managing a Safe Environment
0-7506-3301-8	7. Managing Lawfully - Safety, Health and Environment
0-7506-37064	8. Preventing Accidents
0-7506-3302-6	9. Leading Change

B. Managing Resources

0-7506-3303-4	1. Controlling Physical Resources
0-7506-3304-2	2. Improving Efficiency
0-7506-3305-0	3. Understanding Finance
0-7506-3306-9	4. Working with Budgets
0-7506-3307-7	5. Controlling Costs
0-7506-3308-5	6. Making a Financial Case

C. Managing People

0-7506-3309-3	1. How Organisations Work
0-7506-3310-7	2. Managing with Authority
0-7506-3311-5	3. Leading Your Team
0-7506-3312-3	4. Delegating Effectively
0-7506-3313-1	5. Working in Teams
0-7506-3314-X	6. Motivating People
0-7506-3315-8	7. Securing the Right People
0-7506-3316-6	8. Appraising Performance
0-7506-3317-4	9. Planning Training and Development
0-75063318-2	10. Delivering Training
0-7506-3320-4	11. Managing Lawfully - People and Employment
0-7506-3321-2	12. Commitment to Equality
0-7506-3322-0	13. Becoming More Effective
0-7506-3323-9	14. Managing Tough Times
0-7506-3324-7	15. Managing Time

D. Managing Information

0-7506-3325-5	1. Collecting Information
0-7506-3326-3	2. Storing and Retrieving Information
0-7506-3327-1	3. Information in Management
0-7506-3328-X	4. Communication in Management
0-7506-3329-8	5. Listening and Speaking
0-7506-3330-1	6. Communicating in Groups
0-7506-3331-X	7. Writing Effectively
0-7506-3332-8	8. Project and Report Writing
0-7506-3333-6	9. Making and Taking Decisions
0-7506-3334-4	10. Solving Problems

SUPER SERIES 3 USER GUIDE + SUPPORT GUIDE

0-7506-37056	1. User Guide
0-7506-37048	2. Support Guide

SUPER SERIES 3 CASSETTE TITLES

0-7506-3707-2	1. Complete Cassette Pack
0-7506-3711-0	2. Reaching Decisions
0-7506-3712-9	3. Managing the Bottom Line
0-7506-3710-2	4. Customers Count
0-7506-3709-9	5. Being the Best
0-7506-3708-0	6. Working Together

To Order - phone us direct for prices and availability details
(please quote ISBNs when ordering)
College orders: 01865 314333 • Account holders: 01865 314301
Individual purchases: 01865 314627 (please have credit card details ready)

We Need Your Views

We really need your views in order to make the Super Series 3 (SS3) an even better learning tool for you. Please take time out to complete and return this questionnaire to Tessa Gingell, Pergamon Open Learning, Linacre House, Jordan Hill, Oxford, OX2 8BR.

Name : ....

Address : ....

....

Company & Position (if applicable) : ....

Title of workbook : ....

If applicable, please state which qualification you are studying for. If not, please describe what study you are undertaking, and with which organisation or college:

....

Please grade the following out of 10 (10 being extremely good, 0 being extremely poor):

Content   .... Appropriateness to your position   ....

Readability   .... Qualification coverage   ....

What did you particularly like about this workbook?

Are there any features you disliked about this workbook? Please identify them.

Are there any errors we have missed? If so, please state page number: ....

How are you using the material? For example, as an open learning course, as a reference resource, as a training resource etc.

....

How did you hear about Super Series 3?:

Word of mouth: Through my tutor/trainer: Mailshot:

Other (please give details): ....

Many thanks for your help in returning this form.